LOW-LEVEL RADIOACTIVE WASTE

From Cradle to Grave

Edward L. Gershey

Robert C. Klein

Esmeralda Party

Amy Wilkerson

VNR VAN NOSTRAND REINHOLD
New York

AUG 2 9 1991

Copyright © 1990 by Van Nostrand Reinhold

Library of Congress Catalog Number 89-48403

ISBN 0-442-23958-0

Printed in the United States of America

Van Nostrand Reinhold
115 Fifth Avenue
New York, New York 10003

Van Nostrand Reinhold International Company Limited
11 New Fetter Lane
London EC4P 4EE, England

Van Nostrand Reinhold
480 La Trobe Street
Melbourne, Victoria 3000, Australia

Nelson Canada
1120 Birchmount Road
Scarborough, Ontario M1K 5G4, Canada

16 15 14 13 12 11 10 9 8 7 6 5 4 3 2 1

Library of Congress Cataloging-in-Publication Data

Gershey, Edward L.
  Low-level radioactive waste: from cradle to grave /
Edward L. Gershey, Robert C. Klein, Esmeralda Party, Amy Wilkerson.
    p. cm.
  Includes bibliographical references.
  ISBN 0-442-23958-0
  1. Radioactive wastes. 2. Low-level radiation—Environmental
aspects. I. Gershey, Edward L.
TD898.L665 1990   89-48403
363.72'89—dc20   AAX4043   CIP

CONTENTS

PREFACE

The subject of low-level radioactive waste (LLRW) is important, not only because of the relationship between our standard of living and waste production but also because society's response to this issue is indicative of how we will respond to the larger waste crises—solid, hazardous, infectious, and agricultural—facing the nation. It is important to learn from the attempts to meet federally imposed deadlines and site new LLRW disposal facilities.

In this book we tell the story of LLRW: what it is, where it comes from, how and where it is disposed of, the history of its disposal, what management and disposal alternatives are available, and what plans have been made for the future. We also address the risks associated with LLRW and how our society has responded thus far to this and related issues. Although we present this story in an integrated format, each of the chapters can be read independently as an essay with documentation on a specific subtopic. The LLRW issue is technically uncomplicated and understanding does not require expertise in health physics or radiobiology. However, to aid the reader we have included a glossary, extensive references, and recommendations for additional reading.

This book should be of value to the legislators, regulators, and educators confronting waste disposal problems as well as to health professionals, environmental groups, journalists, media specialists, and those chosen by government to answer questions posed by the affected communities. Establishing a common base of factual information is essential to communicating both the concerns about and solutions to waste disposal problems.

ACKNOWLEDGMENTS

The authors are grateful to Kerry Bennert and John Hsu (DuPont-New England Nuclear Inc.), Jay Dunkleberger (NYS LLRW Siting Commission), Ronald Fuchs (EG&G Idaho), David Kolasinksi (Amersham, Inc.), David MacKenzie (Brookhaven National Laboratory), John Matuszek (NYS Health Department), Alan Jones and Peter Pastorelle (NDL Organization, Inc.), Michael Spall (Con Edison, Indian Point), and Jack Spath (NYSERDA) for information and clarification of data and practices. We are especially grateful to J. Matuszek for not allowing us to be too "even-handed." Edward L. Gershey also appreciates the encouragement of the New York Academy of Medicine's Working Group on Radiation and Health, especially David Becker, John Laughlin, Letty Lutzker, Christopher Marshall, Jeanne St. Germain, and Rosalyn Yalow.

INTRODUCTION

Life has evolved in the omnipresence of radiation. Mutations, some of which are caused by radiation, provide a source of biological variation upon which the forces of natural selection act. This is the molecular basis of evolution. Most experts focus on radiation's ability to induce cancer, and indeed more is known about the carcinogenic potential of radiation than any other toxic substance. Adequate information about health hazards is known for only about 10% of the 63,000 chemicals in common use, whereas those from radioisotopes have been more thoroughly investigated. Likewise, we have more experience with managing and disposing of radioactive materials than any other hazardous substance. Although there are approximately 30,000 hazardous waste sites listed by the U.S. Environmental Protection Agency (EPA), there are only six commercial radioactive waste disposal sites. The LLRW sites have been well studied and characterized, and not one is on the National Priority List of about 1000 sites requiring remedial cleanup (Upton et al., 1989). In comparison with other wastes, radioactive waste is unusual in that the very property that makes the waste hazardous, its radioactivity, will disappear with time. Nonetheless, radioactive waste will remain, some of it for scores of generations. Radioactive waste disposal is just one of the growing number of environmental challenges that face our society. If we cannot learn from our past failures in confronting the LLRW issue, which already has technical solutions, we will be doomed to fail with the others that are now emerging.

Most people think of radiation as rays with great destructive potential that we cannot hear, see, smell, taste, or touch. They seem unaware that selective use of radiation can be beneficial. In fact, radioactive materials are utilized in a broad range of activities that have become essential to our quality of life. Public awareness of these applications varies. Some uses, such as in smoke detectors and other measuring devices, are of low profile. Although most people are aware of nuclear power reactors, few know of the millions of medical diagnostic and treatment applications performed annually to combat scores of diseases and which serve as the basis for future advances in medicine. Even fewer realize that trace amounts of radiochemicals offer improved sensitivity over other chemical methods of analysis and are integral to research in nearly all fields of modern science. Equally unrecognized is the dependency of these applications upon the manufacture of radiochemicals and hence upon common disposal solutions. Radioactivity is used extensively in nondestructive testing of welds and piping, irradiation of food, and myriad other applications. Radioactive waste is an unavoidable by-product from all of these applications and remains a focus of public concern.

Although some individuals argue that the best way to deal with waste is simply not to produce it, few are willing to sacrifice the benefits associated with the production of these wastes. The middle-of-the-road position, currently in vogue, is simply to reduce the amount of waste produced. When radiation is involved, the public response is inconsistent. For example, while there is a growing concern that continued use of fossil fuels for electricity is environmentally unsound, the United States has been unwilling to accept that nuclear energy is an environmentally cleaner alternative (Houk, 1989). It is even cleaner than the collection and use of solar energy (Cohen, 1986). The high-tech manufacture of photovoltaic cells can lead to ground-water contamination from toxic chemicals and other environmental problems associated with huge arrays of solar energy–collecting panels. We assume that, for the near future, our society will continue to rely upon radioactive materials for many purposes. Not only our life-style but also our health relies upon the use of radioactive materials and ultimately upon their safe disposal. Therefore we will have to become more knowledgeable about radioactive waste, what it is, where it comes from, and how it can be reduced, treated and disposed of.

The subject of this book is LLRW, not high-level radioactive waste (HLW). With few exceptions, notably the spent fuel from nuclear reactors, all of the radioactive waste produced by biomedical and academic institutions, industry, and nuclear power generators is LLRW. We have confined our analysis to the waste by-products of the nonmilitary applications of radiation. Reliable information is available only from nonmilitary generators, and it is this area over which we, the public, can gain understanding and have a more direct impact on the decisions that affect the quality of our lives.

Waste disposal strategies cannot succeed if they are subject to change on a four-year political cycle. Too often, political responses have been characterized by postponement and deferment. Regulatory responsibilities have been divided among many different agencies at local, state, and federal levels. In 1980, an act of Congress (Public Law 96-573, 1980) shifted the responsibility for waste disposal from the federal government to the states. During the following five years little progress occurred. The Amendments Act of 1985 (Public Law 99-240, 1985) set deadlines with assessments through 1993 in an effort to hasten a response. Individual states, or groups of states called "compacts," are now involved in meeting their mandated obligations to avoid penalties. Multiple localities must be persuaded to accept treatment and disposal sites, which to date few have been willing to do. As a result of decentralization, states and localities across the country are still in the process of determining how to manage disposal of their LLRW. The time for finding LLRW disposal solutions is running short.

Fortunately the LLRW problem can be solved. Only one exposure, insignificant to the public, has resulted from LLRW (see section on Maxey Flats in Chapter 2). This exposure had no health consequences. Potential exposures to the public from LLRW are so low that they are indistinguishable from the fluctuations of naturally occurring radiation. LLRW presents minimal risk in comparison with risks in other industries and voluntary activities. The problem is that the disposal of LLRW and the fear of radiation, which translates into demands for zero exposure, are irreconcilable. We believe that current disposal methods and volume-reduction measures are capable of meeting the disposal challenge. We describe and discuss them in terms of past experience with their financial, environmental, and health impacts. Policies regulating this technology must be clear enough to follow without doubt or distrust.

Advocates of public health and those knowledgeable of hazardous waste issues must lead us to recognize that the benefits are precariously dependent upon acceptance of the small risks associated with very low levels of radioactivity. Hesitation or equivocation on radiation issues by health and safety professionals or public officials sends strong signals to a distrustful, apprehensive public. Also, phobic perceptions of radiation are heavily influenced by a sensationalizing press that has difficulty evaluating and communicating technical information.

Unfortunately, the secrecy and deliberate misinformation with which the federal government has managed HLW and defense waste has contributed to public mistrust. Nonetheless, decisions must be made.

Who will make these decisions? How will those who wish to participate in the decision-making process be sufficiently educated to ensure that their decisions will be in the public's interest? Individuals must either take the time to understand technical issues or be willing to trust the judgment of those who do. The broader the base of informed citizenry, the greater the likelihood of finding informed leadership. Ultimately, it is the level of literacy that will determine our society's course and humanity's future.

Society's Awareness of Radiation

Although the human experience with radiation has its origins in antiquity, the relatively recent experience with man-made radiation has had the greatest impact upon our regulatory policies. Soil and rocks, air, water, building materials, and food constitute the major sources of radiation exposure to the public. Natural environmental sources account for 82% of the average annual dose, 55% coming from radon (BEIR, 1980; BEIR, 1988; NCRP, 1987) (Figure I-1). The residual radioactivity from weapons testing is the largest source of man-made atmospheric radiation. However, 15% of the dose we annually receive is from medical procedures. The radioactivity associated with LLRW is small by comparison to that from natural sources and fallout, yet just the mention of "radioactivity" conjures up the dramatic image left by the first military application of nuclear energy, the mushroom clouds erupting over Hiroshima and Nagasaki. These clouds continue to darken the public's perception of radioactivity. A brief chronology of the nuclear era may help to highlight the sources of public information that have contributed to the opinions now shaping U.S. nuclear policies.

1895   Wilhelm Conrad Roentgen discovered radioactivity.

1942   First production of nuclear energy at Fermi reactor.

1945   Bombings of Hiroshima and Nagasaki focused attention on the dangers of nuclear energy.

1946   Atomic Energy Commission (AEC) established civilian control of nuclear energy.

1954   Federal regulations (AEC, revised) permitted private utilities to operate nuclear reactors.

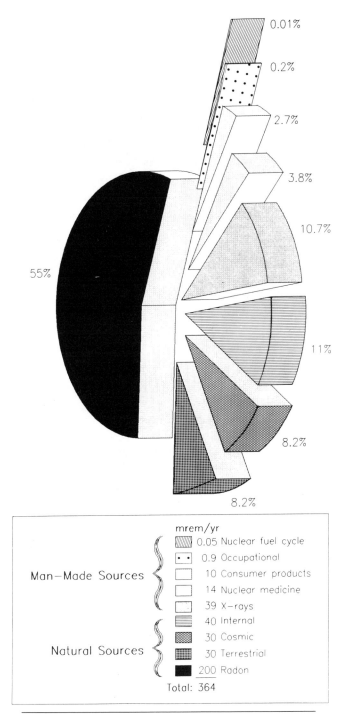

0.01%

0.2%

2.7%

3.8%

10.7%

55%

11%

8.2%

8.2%

mrem/yr		
Man–Made Sources	0.05	Nuclear fuel cycle
	0.9	Occupational
	10	Consumer products
	14	Nuclear medicine
	39	X–rays
Natural Sources	40	Internal
	30	Cosmic
	30	Terrestrial
	200	Radon
Total: 364		

FIGURE I-1. Average annual dose from natural (82%) and man-made (18%) background radiation.

1956   England completed construction of the world's first large-scale
       nuclear power plant.
1962   United States, U.S.S.R., France, and Great Britain banned
       above-ground nuclear testing.
       Opening of first commercial disposal facility at Beatty, Nevada.
1969   National Environment Policy Act set requirements for envi-
       ronmental impact statements.
1970   OSHA set worker exposure standards for X-rays, accelerators,
       natural radioactive materials.
       EPA jurisdiction established radiation protection activities for
       the environment.
       The Clean Air Act favored nuclear energy over fossil fuel.
1972   United States stopped ocean disposal of radioactive waste.
1973   Arab oil embargo raised the price of fossil fuel, focusing at-
       tention on energy.
1974   AEC responsibilities were divided between Energy Research
       Development Administration and NRC.
1976   Resource Conservation and Recovery Act (RCRA) set "cradle-
       to-grave" control of hazardous wastes.
1977   DOE office of Nuclear Waste Management asked to develop
       repositories for defense and HLW.
1979   Temporary closure of the operating commercial LLRW disposal
       sites highlighted the crisis.
       Three Mile Island plant accident: insignificant exposures but
       enormous socioeconomic impact.
1980   The Superfund Act regulated the cleanup of hazardous waste
       sites and spills.
       LLRW Policy Act made states responsible for disposing of
       LLRW via regional facilities.
1981   NRC set de minimis levels for disposal of animal carcasses and
       scintillation fluids.
1984   Superfund amendment placed a moratorium on land disposal
       of certain types of chemicals.
1985   LLRW Policy Amendments Act extended the 1986 deadline to
       1993.
1986   Nuclear power plant accident at Chernobyl, U.S.S.R.

Following the discovery of radioactivity in the late 1800s, it was
recognized by the 1920s that radiologists, radium-dial painters, nuclear
chemists, and miners were developing skin cancers and anemia, and

dying prematurely. Consequently, voluntary radiation protection standards were established and over the years lowered to remain conservative with regard to the epidemiological data.

The Fermi reactor produced the first controlled, sustained yield of nuclear energy in 1942, leading to the development of the atomic bomb. The Atoms for Peace Program followed World War II and placed the commercial use of radioactive materials under regulatory control. The Atomic Energy Commission disposed of radioactive wastes by ocean dumping and shallow land burial. Abundant space and low population densities made land burial practices cost effective. Fortunately, the impacts were small from both of these disposal practices because, at the time, waste management reflected little thought or understanding of potential environmental impacts.

The first power reactors appeared in the late 1950s and, in the United States, multiplied to the point of supplying 10% of the nation's electricity. Biomedical applications of radiation also grew rapidly. Industrial uses of radioactivity expanded to many applications beyond the painting of watch dials with radium.

Commercial burial sites were opened in the 1960s, leading to heightened political and social awareness concerning the environment. Various laws aimed at conserving natural resources were enacted, and a well-developed antinuclear movement acquired political strength.

In the 1970s, the problems associated with landfills and the burial of toxic materials, dramatized by Love Canal, were widely publicized. The United States stopped burial of radioactive waste at sea. Problems were found with existing LLRW sites, three of the six sites closed, and remedial cleanups were begun. In 1979, the three remaining commercial burial sites closed temporarily because the governors of those states wished to highlight their political concerns at being the nation's dumping grounds. The number of power reactors in the United States reached a peak, and LLRW became associated with the hazards of high-level nuclear sources.

In the 1980s, a climate of deregulation and decentralization led the federal government to regard LLRW disposal as a state problem. In contrast, European governments provided solutions through centralized approaches. The United States, with 111 active nuclear power plants, has the most plants and derives 18% of its electricity from nuclear sources; Canada and England derive 15%; Bulgaria, Finland, Hungary, Japan, Spain, Sweden, Switzerland, and West Germany each derive 30%; Taiwan and South Korea 50%; and Belgium and France 65% and 70%, respectively (IAEA, 1989). Since the accident at Three Mile Island, few new nuclear plants have been completed in the United States, plans for many have been canceled, and none have been ordered. The 1986 incident at Chernobyl has reinforced the public's fear of radiation.

References

Cohen, B. L. 1986. A generic probabilistic risk analysis for a high-level waste repository. *Health Physics* 51:519–528.

Committee on the Biological Effects of Ionizing Radiations (BEIR). 1980. *The Effects on Populations of Exposure to Low Levels of Ionizing Radiation: 1980*. Washington, D.C.: National Academy Press.

Committee on the Biological Effects of Ionizing Radiations (BEIR). 1988. *Health Risks from Radon and Other Internally Deposited Alpha Emitters*. BEIR IV Report. Washington, D.C.: National Academy Press.

Houk, V. 1989. Science and society: low level radioactive waste, controversy and resolution. *Bulletin of the New York Academy of Medicine* 65:485.

International Atomic Energy Agency (IAEA). 1989. *Nuclear Power and Fuel Cycle: Status and Trends*. Vienna: IAEA.

National Council on Radiation Protection and Measurements. 1987. *Ionizing Radiation Exposure of the Population of the United States*. NCRP Report No. 93. Bethesda, Maryland: NCRP.

Upton, A. C., T. Kneip, and P. Toniolo. 1989. Public health aspects of toxic chemical disposal sites. *Annual Review of Public Health* 10:1–25.

1 WHAT IS RADIOACTIVE WASTE?

Radioactive waste is not a singular material. Instead, it is generated in diverse forms that have traditionally been distinguished by their source, not by their physical characteristics. Radioactive wastes vary greatly in their chemical and radioactive composition and therefore in their potential for environmental and public health impacts.

In the United States, as well as in most other nations involved in the nuclear arms race, the vast majority of radioactive waste originates as by-products from nuclear weapons production. For example, Department of Energy (DOE) defense-related wastes account for more than 70% of all radioactive wastes generated in the United States (DOE, 1988), despite the fact that the United States has the world's largest civilian nuclear power program. Most of this defense waste has been generated and managed under a shroud of national defense secrecy, without the intense public scrutiny that has marked commercial nuclear activities. This secrecy abruptly changed in the late 1980s with widespread publicity over the DOE's poor past performance in waste management. Although many of these problems originated with the Manhattan Project during World War II, some poor practices have continued to the present. Multi-billion-dollar budgets have been allocated to the task of cataloging, documenting, and cleaning up active and formerly used sites. Remedial action programs and environmental restorations are now under way at most of the larger DOE facilities.

Because of their variety, it is important to understand the differences between the various kinds of radioactive waste before focusing on low-level radioactive waste (LLRW). Table 1-1 lists the sources, amounts, volumes, and relative hazards associated with the different kinds of radioactive waste. The concentration of these wastes is shown graphically in Figure 1-1. Clearly, many orders of magnitude separate

TABLE 1-1. Radioactive Wastes Generated in the United States Through 1988[a]

Waste	Principal Generators	Typical Nuclides	U.S. Inventory Curies[b]	U.S. Inventory m^3	Surface Exposure	Hazard Duration (years)	Overall Hazard Potential
Spent fuel	Nuclear power plants, DOE activities	^{137}Cs, ^{60}Co, ^{235}U, ^{238}U, $^{239\text{-}242}$Pu	1.8×10^{10}	6.80×10^3	High	$>10^5$	Requires isolation in perpetuity
High-level	DOE reprocessing of spent fuels	^{90}Sr-^{90}Y, ^{137}Cs, ^{144}Ce, ^{106}Ru, $^{239\text{-}242}$Pu	1.3×10^9	3.82×10^5	High	$>10^5$	Requires long-term isolation
Transuranic	Plutonium production for nuclear weapons	$^{239\text{-}242}$Pu, ^{241}Am, ^{244}Cm	4.1×10^6	2.80×10^5	Moderate	$>10^5$	Soluble and respirable
Mill tailings	Mining and milling of uranium/thorium ores	^{235}U, ^{230}Th, ^{226}Ra	1.4×10^5	1.20×10^8	Low	$>10^4$	Hazard to worker
Greater than Class C	Nuclear power plants, users and manufacturers of sealed-source devices	^{60}Co, ^{137}Cs, ^{90}Sr, ^{241}Am	2.40×10^6	1.30×10^2	High	500	High
Low-level DOE	Various processes, including decontamination and remedial action cleanup projects	Fission products, ^{235}U, ^{230}Th, α-bearing waste, ^{3}H	1.4×10^7	2.40×10^6	Unknown	$>10^3$	High, poorly managed in the past
Low-level commercial							
Class A	Fuel cycle, power plants, industry, institutions	See Table 1-7	3.6×10^5	1.3×10^6	Low	200	Low
Class B	Principally power plants and industry	"	9.5×10^5	2.7×10^4	Moderate	$\sim10^3$	Moderate
Class C	Power plants, some industry	"	2.5×10^6	6.5×10^3	High	$>10^5$	High

[a] DOE, 1988; OTA, 1988
[b] Ci = 3.7×10^{10} Bq

2

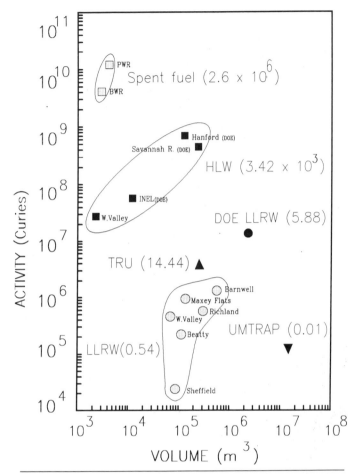

FIGURE 1-1. Activity and volume of radioactive wastes generated in the United States. The average concentration of radioactivity of the wastes is shown in parentheses. Spent reactor fuel has the greatest concentration, followed by HLW, transuranic (TRU) wastes, DOE LLRW, commercial LLRW, and uranium mill tailings. Open symbols are commercially produced wastes, solid ones refer to DOE materials.

the radioactivity in the different classes of waste. Spent fuel and defense high-level radioactive waste (HLW) have the highest concentration and account for most of the activity associated with nuclear wastes. At the opposite end of the spectrum, LLRW and uranium mill tailings have the lowest concentrations but higher volume.

Spent Fuel and High-Level Wastes

Spent fuel is intact nuclear fuel that has been used in a nuclear reactor. It is highly radioactive and poses serious radiation hazards requiring shielding, containment, remote handling, and initially, underwater storage for cooling. Figure 1-2 outlines what is commonly referred to as the nuclear fuel cycle, that is, the processes that mine, fabricate, enrich, use, and dispose of nuclear fuel materials. Although spent fuel contains plutonium and enriched uranium in economically recoverable amounts, commercial fuel reprocessing was discontinued in the United States in 1972 when licensing restrictions led to the closure of the only reprocessing plant (NCRP, 1987). Two other reprocessing facilities were constructed in the 1970s in Illinois and South Carolina but never operated because of government security concerns about plutonium. The DOE, however, continues to reprocess most of its spent fuel. Many other nations with well-developed nuclear power programs, such as France, Germany (FRG), Japan, United Kingdom, and U.S.S.R., are either actively reprocessing or preparing to reprocess their spent fuel (IAEA, 1988a). Reprocessing increases the overall energy efficiency of the nuclear fuel cycle by recycling enriched uranium and recovering plutonium, which can provide a partial fuel substitute for ^{235}U in the so-called mixed-oxide (MOX) and breeder reactors. Negative aspects of reprocessing are the serious potential health consequences from accidental releases of plutonium, which is highly carcinogenic; the use of plutonium in nuclear weapons production; and the possibility that bomb-grade material could end up in the hands of terrorists. Commercial spent fuels are currently stored on-site at nuclear power plants throughout the United States (Figure 1-3).

HLW is generated during the reprocessing of spent reactor fuel (DOE, 1988). Most of the United States inventory of HLW is related to DOE and defense activities and is stored at the Savannah River Plant, South Carolina; Idaho National Engineering Laboratory (INEL), Idaho; and Hanford Reservation, Washington. Although nearly all HLW is from the DOE, a small fraction (<1%) originated from the reprocessing of commercial spent fuels from 1963 to 1972 at the Nuclear Fuel Services Corporation site at West Valley, New York (see Figure 2-1 in Chapter 2).

During spent fuel reprocessing, large volumes of acid and other solvents are used to extract radionuclides chemically from the fuel rod assemblies. These liquids are then treated to precipitate plutonium and uranium and pumped to storage tanks for additional processing.

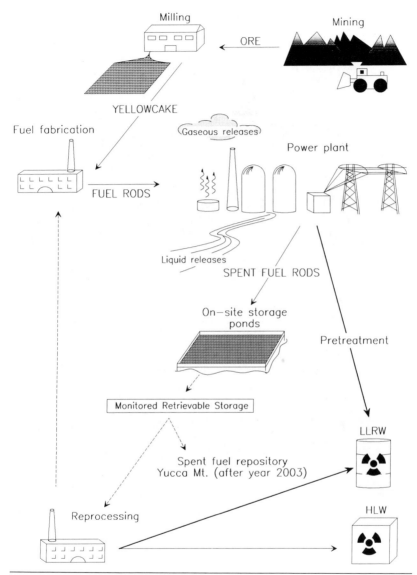

Milling

Mining

ORE

YELLOWCAKE

Fuel fabrication

Gaseous releases

Power plant

FUEL RODS

Liquid releases

SPENT FUEL RODS

On–site storage ponds

Pretreatment

Monitored Retrievable Storage

LLRW

Spent fuel repository Yucca Mt. (after year 2003)

Reprocessing

HLW

FIGURE 1-2. The nuclear fuel cycle. Current practices are identified with solid lines, while those that have either been discontinued or not yet put into operation are shown with dashed lines.

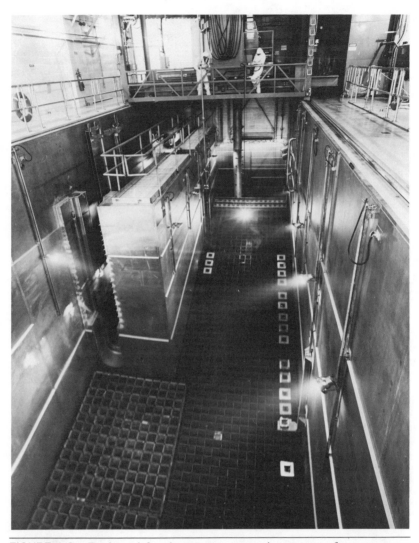

FIGURE 1-3. Pool used for the temporary on-site storage of spent reactor fuel. The fuel rods are held underwater in honeycombed cells as shown along the right-hand wall of the pool bottom. *(Courtesy of Duke Power Company.)*

Disclosure of leaking HLW storage tanks at the Hanford Reservation in eastern Washington helped bring notoriety to the DOE. In the tanks, particulates settle out to form sludges and slurries that must eventually be dewatered. The sludges and precipitates contain high concentrations of radioactive cesium, strontium, yttrium, plutonium, uranium, and other nuclides. Although most of the initial radioactivity in these wastes comes from ^{90}Sr, ^{137}Cs, and other fission products that will decay within the first few hundred years after disposal, HLW and spent fuel will retain hazardous levels of uranium, plutonium, and other actinides for thousands of years. These wastes must be physically isolated from the biosphere. Great attention has naturally been focused on methods to contain them. Like spent fuel, all HLW is stored on an interim basis pending development of a final repository for these wastes.

As part of the Nuclear Waste Policy Act of 1982, Congress mandated that the DOE select and construct one or more repositories for HLW and spent fuel. The research and development of such a repository has a long history, beginning with a National Academy of Sciences recommendation in 1957 that long-term disposal would be best managed by deep geologic burial (NAS, 1957). Initially, the DOE's predecessor concentrated its efforts on evaluating salt-dome formations in the Midwest near Lyons, Kansas, for the burial of spent fuel, HLW, and transuranic wastes. After much politicking, the DOE instead chose the crystalline rock formations at the Yucca Mountain site in Nevada for the non-transuranic wastes. Now in the early stages of characterization of Yucca Mountain, the DOE anticipates that 10 years will be required to fully characterize the site, licensing will take three years, and construction another seven years, giving an optimistic early date for commencement of 2003 (Gertz, 1989). Although the ultimate cost of the project is not known, the initial characterization process alone will cost $1 to 2 billion.

The steadily shrinking storage capacity for spent fuel at commercial nuclear power plants has led to demands for off-site storage of these wastes. This concept, known as "monitored retrievable storage" (MRS), offers temporary off-reactor storage. However, fears that such a facility could open before assurances that final HLW disposal is available, and thus perhaps lengthen the storage period indefinitely, led Congress to link construction of MRS to the licensing phase of the HLW repository (Klein, 1989), effectively postponing MRS for at least another 10 years. Power plant operators have been encouraged by the NRC to begin the necessary expansion of their spent fuel pools. Several power plants have already begun experimenting with methods

FIGURE 1-4.   Fuel rod rack consolidation project. *(Courtesy of Duke Power Company.)*

of fuel rod consolidation to increase the density of their storage ponds (Figure 1-4).

Besides commercial nuclear power plants, some 99 operating and 123 inactive research reactors are located in the United States (IAEA, 1988b). While many are directly or indirectly run by the DOE and other branches of the federal government, industry and academic institutions operate more than half of these reactors. Their impact is small in terms of waste volume, but they provide additional sources of both spent fuel and low-level radioactive waste that must be managed.

Transuranic Waste

Transuranic (TRU) wastes consist of materials contaminated by radionuclides with atomic numbers greater than uranium, such as plutonium, americium, and curium. TRU wastes generally contain less activity but are more voluminous than HLW or spent fuel. They result from every industrial process involving transuranic materials, but are predominantly by-products from the fabrication of plutonium for nuclear weapons (DOE, 1988). In the United States, the DOE is the principal generator of TRU wastes. These wastes pose high health hazards because they tend to be water-soluble, respirable (up to 1% by weight can be less than 10 μm in diameter), and contaminate a variety of physical forms, ranging from unprocessed trash (e.g., absorbent papers, personal protective equipment, plastics, rubber, wood, ion-exchange resins, etc.) to discarded tools and glove boxes. Major producers of TRU waste are the Rocky Flats Arsenal, Colorado; Savannah River Plant, South Carolina; Hanford Reservation, Washington; and Los Alamos National Laboratory, New Mexico. Smaller producers include the Mound Facility, Ohio; Argonne National Laboratory, Illinois; Oak Ridge National Laboratory, Tennessee; and Lawrence Livermore Laboratory, California.

Prior to 1970, the TRU category did not exist. TRU wastes were buried at their production sites in open, unlined trenches and then covered with several meters of earth. At the time of their burial, no plans for the future retrieval of these wastes were made nor has any decision yet been made to exhume the estimated 150,000 m^3 of previously buried TRU wastes. In 1970, the U.S. Atomic Energy Commission (AEC), the NRC's predecessor, adopted a policy requiring that wastes contaminated with a concentration greater than 10 nCi/g (370 Bq/g) of alpha particles be packaged, stored, and disposed of separately from other radioactive wastes; this limit was raised to 100 nCi/g (3700 Bq/g) in 1983. Because of limited storage space at several of the major producing facilities, TRU waste has been shipped to INEL in Idaho Falls, Idaho, since 1970. After its redefinition in 1983, most of the unregulated TRU-containing wastes [i.e., <100 nCi/g (3700 Bq/g)] have been shipped to the Nevada Test Site for disposal by shallow land burial. Approximately 57,000 m^3 of regulated TRU waste (>100 nCi/g or 3700 Bq/g) is currently stored on a temporary, retrievable basis at INEL.

Final disposal of TRU wastes will occur at the Waste Isolation Pilot Plant (WIPP) constructed near Carlsbad, New Mexico (Neill and Chaturvedi, 1989). Attention was focused on this location after the government abandoned prospects for deep geologic disposal at the Lyons, Kansas, site. Although the salt-dome formations at the Lyons site had received favorable review for use in HLW disposal, investigations eventually determined that earlier exploratory petroleum drilling in the area could predispose the site to water infiltration and subsequent radionuclide migration. Political considerations were also a factor. Geologic characterizations at the WIPP began in 1974 under the auspices of the United States Geologic Survey and Sandia National Laboratory. After many changes and revisions to its mission, the WIPP is now designed exclusively for the disposal of TRU wastes. The actual repository lies at a depth of 855 m below the surface and consists of galleries and tunnels mined out of a 600-m-thick salt formation. Capacity is estimated at about 185,000 m^3, for a total inventory of 14 MCi (5.18×10^{17} Bq). Most of these wastes will be moved without special equipment, although provisions are being made for the remote manipulation of the small volume of waste that produces high external surface exposures. Under Public Law 96-164, the WIPP was designated as a research and development demonstration project, exempt from Nuclear Regulatory Commission requirements and those of the National Environmental Policy Act. Nonetheless, TRU wastes will be regulated by both the NRC and EPA since 60% of the waste qualifies as mixed waste (Jones, 1989) (see "Mixed Waste" section, below). Actual construction at the site began in 1983. Early test boreholes showed that the initial site was overlain by a pressurized brine reservoir, which led the DOE to relocate the facility several kilometers to its present site. Limiting brine inflow is of paramount importance to the ultimate success of the WIPP. The volume of brine still expected to enter the facility has apparently not tempered the DOE's decision to begin receiving waste in early 1990.

Transportation of TRU waste to the WIPP will be made using a truck-mounted cask system designed by Westinghouse (Halverson, 1989). This system, called TRUPACT II, uses double-layered stainless steel containers that can be either manually or remotely filled and later unpacked. TRUPACT II has undergone rigorous accident testing and received NRC approval. A fleet of about 50 radar-tracked trailers is planned.

Mining and Milling Wastes

The process of mining and milling uranium and thorium ores generates large quantities of rock, sludge, and liquids. These wastes are contaminated to varying degrees with the parent ore as well as daughter nuclides such as radium, polonium, bismuth, and lead (IAEA, 1987). Wastes are generated during the exploratory and operational phases of mining and consist of large amounts of rock from excavations and liquids from surface drainage, seepage, and in situ leaching. During mine operation, liquid and airborne effluents bearing gases and dusts constitute a significant hazard to both workers and the public. Once high-grade ore is excavated, a typical milling operation (Figure 1-5) involves the chemical extraction or leaching of radioactive minerals from the ore. Heap piles are built and solutions of acids and solvents are recirculated through the pile until extraction yields of acceptable quality are achieved. Leachates are then shunted to evaporators which

FIGURE 1-5. Uranium milling facility at Ambrosia Lake, New Mexico. The barren liquor impoundment is visible beyond the cluster of buildings. *(Courtesy of the Uranium Mill Tailings Project Office, U.S. DOE.)*

further purify the crude product by roasting the concentrate. This dried form is known as yellowcake. Large volumes of liquid containing acids, their salts, heavy metals, organic solvents, and residual radio-nuclides are then pumped to tailings impoundments for settling and decantation of the barren liquor. Unless well controlled, these im-poundments will contaminate local ground and surface water by runoff and seepage.

Once mining is completed, radon from the decay of ^{226}Ra is gen-erally considered the most serious potential health hazard, particularly if the tailings are misused as building materials or fill. Windborne dust can also pose a significant long-term, off-site hazard. Although most mining and milling activities have occurred in sparsely populated areas of the Western states, further processing stages occur throughout the United States. Annual individual doses from mining and milling are small, but collectively they provide the major source of the public's radiation impact from the nuclear fuel cycle (UNSCEAR, 1977; NCRP, 1987).

Mining and milling wastes were poorly managed in the past, re-sulting in the need for federally sponsored programs to upgrade earlier disposal sites. Under the DOE, the Uranium Mill Tailings Remedial Action Program (UMTRAP) will ultimately address the decontami-nation of 24 inactive processing sites and about 4500 adjacent prop-erties (Figure 1-6). Eleven of these sites will be stabilized in situ using concrete and geologic barriers; the remainder will be decontaminated by the removal of soil, equipment, and structures (Caldwell et al., 1989). These wastes are generally disposed of as low-level radioactive waste, but are defined in separate regulations. The decontamination goals of UMTRAP operations are regulated under authority of the EPA (Title 40, Code of Federal Regulations, Part 192, i.e., 40 CFR Part 192).

Low-Level Wastes

As mentioned earlier, radioactive wastes are defined by their source, not their characteristics. The definitions for HLW, TRU waste, and LLRW apply only to wastes regulated under the Atomic Energy Act of 1954 (i.e., wastes from the production and use of source, special nuclear, and by-product materials). In contrast to HLW, LLRW ac-counts for less than 1% of the radioactivity but nearly 85% of the volume of all radioactive wastes (DOE, 1988). LLRW is defined in 10 CFR Part 61 as radioactive waste that does not qualify as any of the wastes described in the preceding paragraphs. For regulatory purposes, LLRW is subdivided into three classes, A, B, and C, depending upon the concentration, energy levels, half-life, and the sources of the radionuclides present in the waste (Table 1-2). Class A wastes contain the lowest amount of radioactivity, followed by Class B and C wastes. The longer-lived nuclides are regulated with far greater stringency than short-lived ones. With the exception of the alpha emitter, Curium 242, all of the individually listed nuclides have half-lives greater than 5 years and five of them are in excess of 1000 years. Nuclides with half-lives equal to or shorter than that of tritium (^{3}H) (12.3 years) will decay to inconsequential levels of radioactivity within 100 to 200 years. Because these nuclides will decay in such a relatively short period of time, they have no upper limits as Class B waste.

The designation of material as LLRW does not necessarily imply low hazard. Radiation hazard is a function of radionuclide concentration, half-life, emission type (mode of decay) and energy, level of protection, and, ultimately, mobility, both through the environment and the body. Figure 1-7a shows the decay of radioactivity with time. The rate of this decay is different for each radioisotope (see the table on the inside cover), and is measured in half-lives, or the amount of time required for half of the radioactive atoms to decay. After 10 half-lives, which may require seconds or millions of years, 0.1% of the initial radioactivity will remain. The mode of decay, that is, the type of particle emitted, and its energy (Figure 1-7b), also greatly affects the hazard of a material. For example, irradiation from gamma rays may penetrate deeply while alpha particles will be easily stopped but discharge their energy within a very short distance. The mode of decay affects the type of protection necessary (Figure 1-7c). The amount and type of radiation we can absorb is a function of the duration of exposure and distance from the exposure source. Radiation exposure decreases with distance according to the inverse square law. Decreasing the time and increasing the distance near a radiation source can greatly lower risk. Shielding is also very effective as a means of providing protection from radiation. Depending on the type of radiation, very little shielding to thick layers of lead or concrete may be required. All persons working with or around radioactive materials will require some form of shielding.

Radiation hazard is, ultimately, a function of the radionuclide's mobility both through the environment and the body (Figure 1-7d, e). For example, 1 Ci (3.7 × 10^{10} Bq) of ^{3}H is an inconsequential external hazard in relation to ^{137}Cs, but the greater mobility of ^{3}H can make it much more of a hazard if ingested or inhaled. Mobility, in a broad sense, refers to the solubility and volatility of an isotope and its chemically bound forms. Interactions between the isotope and the

a

b

local environment can lead to air- and waterborne releases. Like certain pesticides and other organic chemicals, some radionuclides can be concentrated in the food chain, leading to high exposures to humans (e.g., ^{90}Sr). Radionuclides that are not bioaccumulated may instead pass through the soil and contaminate groundwater, or, if volatile, accumulate in the atmosphere. Similar dynamic processes occur in water, where the ocean is a large reservoir of radionuclides.

If ingested, inhaled, or otherwise introduced into the body, radionuclides show differential affinity within the body. For example, iodine will accumulate in the thyroid gland, while strontium and potassium will migrate to bone. Once deposited, radioactive decay products in intimate contact with tissue will lead to possible DNA damage and/ or cell death. The hazard of internally deposited radionuclides is a function not only of radioactive decay but also of metabolism and elimination of the compounds to which they are bound.

The NRC has attempted to incorporate the relative exposure risks from radionuclides into their classification system by regulating LLRW

c

FIGURE 1-6. (*a; left top*) Before, (*b; left bottom*) during, and (*c; above*) after views of the Canonsburg, Pennsylvania, Uranium Mill Tailings Remedial Action Project. (*Courtesy of the Uranium Mill Tailings Project Office, U.S. DOE.*)

TABLE 1-2. Maximum Concentration Limits for Low-Level
Radioactive Waste from 10 CFR Part 61

Radionuclide	Half-life (years)	Maximum Concentration Limits (Ci/m³)[a]		
		Class A	Class B	Class C
Nuclides with half-lives <5 years[b]	<5.0	700.000	NL[c]	...[d]
⁶⁰Co	5.3	700.000	NL	...
³H	12.3	40.000	NL	...
⁹⁰Sr	28.0	0.040	150.0	7000.00
¹³⁷Cs	30.0	1.000	44.0	4600.00
⁶³Ni	92.0	3.500	70.0	700.00
⁶³Ni in activated metal	92.0	35.000	700.0	7000.00
¹⁴C	5,730.0	0.800	...	8.00
¹⁴C in activated metal	5,730.0	8.000	...	80.00
⁹⁴Nb in activated metal	20,000.0	0.020	...	0.20
⁵⁹Ni in activated metal	80,000.0	22.000	...	220.00
⁹⁹Tc	212,000.0	0.300	...	3.00
¹²⁹I	17,000,000.0	0.008	...	0.08
α-emitting transuranic nuclides with half-lives <5 years	<5.0	10.000 nCi/g	...	100.00 nCi/g
²⁴²Cm	0.45	2,000.000 nCi/g	...	20,000.00 nCi/g
²⁴¹Pu	13.2	350.000 nCi/g	...	3,500.00 nCi/g

[a] 1 Ci = 3.7×10^{10} Bq
[b] Including, but not limited to: ³²P, ³⁵S, ⁵¹Cr, ⁵⁴Mn, ⁵⁵Fe, ⁵⁸Co, ⁵⁹Fe, ⁶⁵Zn, ⁶⁷Ga, ¹²⁵I, ¹³¹I, ¹³⁴Cs, ¹⁴⁴Ce, and ¹⁹²Ir.
[c] No upper limit on concentration.
[d] Class not defined for these nuclides.

primarily according to half-life and mobility. Some LLRW can produce high external radiation exposures that require shielding and administrative control during transport and handling. Most of the materials disposed of as LLRW are contaminated with beta- and gamma-emitting nuclides. A small fraction is contaminated with alpha-emitting nuclides whose higher hazard is reflected in very low permissible concentration limits (Table 1-2).

All classes of LLRW must meet the same minimum form and packaging requirements prior to disposal, with additional requirements set for Classes B and C. Wastes cannot be packaged in cardboard containers, liquids must be solidified or packed with sufficient material to absorb twice the volume of the liquid waste, and solid wastes must

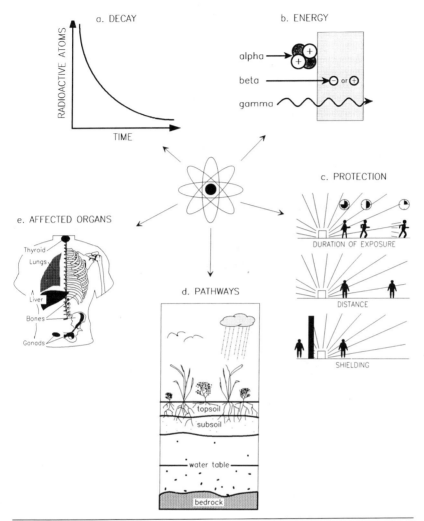

FIGURE 1-7. The hazard from any radionuclide is a function of its mode of decay and energy, mobility and environmental fate, intake and interaction within the body, and half-life. (*a*) Decay of radioactivity with time. (*b*) The type and amount of energy released during radioactive decay will determine how far the radiation can penetrate and how much energy it can impart. (*c*) Proper external radiation protection must take into account duration, distance, and shielding. (*d*) Pathways of radioisotopes through the biosphere. (*e*) Different radionuclides have affinity for different parts of the body.

contain less than 1% free-standing liquid. As handled today, most wastes are packaged into low-carbon steel cylindrical drums, generally of 55-gal (209-L) capacity. The waste may receive pretreatment (see Chapter 3) and is then shipped via brokers for disposal in one of the three operating shallow land burial sites (Figure 1-8). Other requirements, such as the inactivation of pathogenic materials and the absence of hazardous chemicals, must also be met. These minimum requirements are intended to facilitate LLRW handling and provide protection for personnel at the site. Class B and C wastes contain higher levels of radioactivity than Class A materials and must therefore meet more

FIGURE 1-8. Preparing drums of low-level radioactive waste for burial at the Richland, Washington, disposal facility. *(Courtesy of American Ecology.)*

rigorous stability criteria in order to achieve greater structural integrity. Protection from inadvertent intrusion by humans or animals at the disposal site must also be provided when disposing of Class C wastes.

A valuable perspective on commercial LLRW is gained by analyzing the facilities that generate the wastes. These basically fall into four major groups: utilities with nuclear power plants, industrial, institutional (clinical, academic, and biomedical research), and nonmilitary government sources. Table 1-3 lists the kinds of generator facilities and some general characteristics of their waste. Over the past decade numerous studies (NLLRWMP, 1980–1989; Kempf, 1985; Bowerman et al., 1986; MacKenzie, 1988b) have attempted to examine LLRW in terms of these groupings of generators. Although subject to reporting errors and inconsistencies by the generators, waste brokers, and disposal sites, these studies provide the most comprehensive and readily available public data on the subject. Prior to about 1985, national data are very limited since only utility generators were described in detail.

TABLE 1-3. Principal Forms of Low-Level Radioactive Waste Shipped for Disposal[a]

Generator	Facilities Included	Principal Forms of Waste		Volume (%)
Fuel cycle	Nuclear power plants, fuel fabricators & processors, research & development	Dry compressible waste		56
		Stabilized liquids		41
		Other		3
			Total	100
Industrial[b]	Manufacturing, nondestructive testing, mining, research & development	Dry compressible waste		40
		Stabilized liquids		10
		Depleted source material		44
		Other		6
			Total	100
Institutional[b]	Hospitals & clinics, research institutions, private practices	Dry compressible waste		92
		Stabilized liquids		4
		Biohazardous waste		4
			Total	100
Government[b]	Federal, state, & local agencies	Dry compressible waste		93
		Other		7
			Total	100

[a] Ko, 1988; NLLRWMP, 1988

[b] Although liquid scintillation wastes can contribute up to 40% of the volume of the wastes generated in these categories, deregulation has removed this waste from disposal as LLRW.

Utility and Fuel-Cycle Generators

Fuel-cycle and utility waste is probably described more accurately than other sources since the nuclear power industry has been under intense financial and legislative pressure to reduce their LLRW volume. Much of this LLRW consists of compacted trash and dry wastes, filters, and tools. These wastes may contain up to 1% freestanding liquid. Unlike other generators, a significant volume of utility waste is initially produced in liquid or slurry forms (wet wastes) that must be dewatered or otherwise solidified to contain less than 0.5% free-standing liquid before disposal. These wet wastes are derived from the decontamination of reactor cooling water and are made up largely of spent ion-exchange resins and evaporator sludges. Currently, over 100 nuclear power plants (Figure 1-9) are operating in the United States and their waste production is strongly related to their continued operational and refueling cycles. Since virtually all of these plants were constructed in the 20-year period prior to 1979, and their designs limit their active life to 30 to 40 years (IAEA, 1988a), by the year 2015 nearly all of them will be shut down. Future LLRW volumes will be greatly influenced by the decommissioning of these reactors.

No new orders have been placed for nuclear power plants since the 1978 accident at the Three Mile Island plant in Harrisburg, Pennsylvania. Although no members of the public were affected by radioactivity from the accident and only a few workers at the plant received more than their routine occupational dose, recent reports confirmed that the fuel rod assembly underwent greater damage than first indicated (Booth, 1987). Although the accident might have been avoided, the experience graphically demonstrated the efficacy of the emergency core-cooling system. Many more safeguards and backup systems than before this accident are now required in order for a reactor to receive an operating license from the NRC, yet the future of nuclear power in the United States remains uncertain. Surprising to many opponents of nuclear power, since 1945 there have been only 29 deaths directly attributable to radiation accidents at nuclear power plants and all of these occurred at Chernobyl (IAEA, 1988c). Other than those from nuclear facilities, the greatest number of radiation-related deaths have come from accidents involving high-activity radiation sources such as medical irradiators and sterilizers (17 deaths). Of the various technologies for producing electricity, at least operationally, fossil fuel sources are the most dangerous, followed in order by steam, hydropower, and nuclear in terms of expected mortality to workers and the public (CSA, 1989; Houk, 1989; Schurr et al., 1979). Due to naturally

FIGURE 1-9. A nuclear power reactor. The large steam towers in the foreground are cooling water heat exchangers. The reactor building is at the lower right-hand corner. *(Reprinted from the EPRI Slide Library Catalog with permission of the Electric Power Research Institute.)*

occurring radon in coal, gaseous effluents from coal-fired plants produce radiation doses to the public of about the same or greater magnitude as nuclear power plants (McBride et al., 1978; Cohen, 1986). Three million times more energy per unit weight is derived from using uranium or plutonium in a nuclear reactor than from burning natural gas, oil, or coal. This greater efficiency translates into great reductions in chemical pollution and waste generation over traditional methods for producing electricity. Plans to develop smaller, standardized, inherently safe reactors (Taylor, 1989b) may renew interest in nuclear power as worries over growing dependence on foreign oil and environmental problems associated with coal ash disposal and global levels of carbon dioxide move closer to the public's eye.

Industrial Generators

Industrial LLRW is highly process-specific and therefore difficult to characterize. Although there are over 4000 industrial LLRW generators, fewer than 1% have a significant impact on the overall volume of LLRW. The principal industrial generators are manufacturing companies that produce radioactive materials for fuel- and non-fuel-cycle use. Manufacturing companies produce large amounts of compacted dry waste, metals, solidified evaporator bottoms, and contaminated personal protective equipment. Chief among these firms are Cintichem (a subsidiary of Hoffmann-LaRoche), DuPont-New England Nuclear, Amersham Corporation, ICN Radiochemicals, Sigma Radiochemicals (formerly Pathfinder Laboratories), General Electric Company, American Radiolabeled Chemicals, Inc., and Teledyne Isotopes (MacKenzie, 1988a,b). Most of their short-lived radionuclides are generated in-house by cyclotrons and mini-reactors, whereas ^3H and ^{14}C are usually purchased from the DOE or other agencies such as Atomic Energy of Canada. Although listed as industrial generators, much of the waste from these companies comes from the production of radioactive chemicals for agricultural, environmental, and biomedical research and clinical procedures. A few of these generators (e.g., radiochemical companies) produce most of the ^3H- and ^{14}C-contaminated wastes, predominantly packaged as Class B waste (MacKenzie, 1988b). In addition to radiochemicals, manufacturing companies also produce a variety of consumer products that contain radioactivity such as smoke detectors, luminous and self-illuminated devices, static eliminators, and moisture gauges. A handful of manufacturing companies fabricate nuclear fuel for reactors and equipment with high-activity sources. Other sources of industrial LLRW are corporate research and development laboratories, nondestructive testing services, and irradiators used for sterilization (Figure 1-10). Radioactive sources are also integral to many devices, such as those used to measure liquid levels in cans, moisture and permeability of soil and rock, particles (smoke), and thickness of materials. Corporate research and development laboratories produce a small fraction of the industrial LLRW and have a waste profile very similar to institutional LLRW. If the

FIGURE 1-10. Instrument using gamma radiation to test the integrity of welded pipe joints. (*Reprinted from the EPRI Slide Library Catalog with permission of the Electric Power Research Institute.*)

national surveys classified LLRW by content and end-product user instead of their generating sources, a large portion of this industrial waste would in fact be considered institutional.

In the past, service companies that decontaminate and decommission reactors and large manufacturing plants, and LLRW brokers who collect and dispose of wastes from numerous smaller-volume generators were treated as if they themselves had generated the wastes. National waste surveys that provide a breakdown of LLRW by generator category often combined the wastes from these companies with manufacturers. This led to erroneously high reports of industrial volumes of waste. This problem was magnified by also including the numbers from the original generators. Like manufacturers, service companies produce large amounts of compacted dry wastes and trash, but they also generate irradiated and contaminated reactor and plant components, many of which are too large to be disposed of in standard 55-gal drums. Some of the larger service companies include US Ecology, Babcock and Wilcox, Adco, General Atomics, Westinghouse, and Nuclear Fuel Services. Brokers handle diverse waste streams. Their waste streams are varied but consist mostly of compacted dry wastes from institutional, industrial, and clinical sources. Reliance on brokers for hauling wastes, however, has made accurate reporting of waste statistics difficult since, at least in the past, the original generator has often not been identified on the accompanying waste manifest. Questions over who really generated the wastes were addressed by some recent changes made in waste manifest requirements by the NRC in 1987, which now allow for tracking back to the original generator.

Institutional Generators

Institutions are the most numerous of all LLRW generators. Institutions include colleges, universities, nonprofit research organizations, hospitals, and medical clinics. Radioactive materials are ubiquitous throughout all academic and biomedical research and educational activities. Their use is invaluable as radioactivity greatly improves sensitivity and specificity over chemical assay methods of detection (Figure 1-11). In universities radioisotopes are used to aid classroom and laboratory instruction in physics, biology, chemistry. geology, and ecology. Research in modern biology and medicine hinges upon the continued and probably expanded use of small amounts (micro- to millicurie or 10 kilo- to megabecquerel) of radioactivity in tracer experiments. Furthermore, radiolabeled chemicals are essential for predicting the field behavior of pesticides and other environmentally released compounds. Similarly, most new drug formulations undergo metabolic studies that use radioactive compounds.

Institutional LLRW is characterized by radioactively contaminated liquid scintillation waste and large amounts of trash (including personal protective equipment and glass and plastic laboratory supplies), animal carcasses, and pathological and chemical wastes. Prior to their deregulation in 1981, liquid scintillation fluids and animal carcasses accounted for about 50% of the institutional LLRW volume (Roche-Farmer, 1980). The absence of well-developed waste management programs at most institutions has caused large volumes of waste with extremely low radioactivity to be shipped for burial in the same manner as higher-activity wastes. Recommendations such as those by Party and Gershey (1989) and the NLLRWMP (1987b) will have a significant impact on LLRW volume if adopted by more generators.

The clinical use of radioisotopes for diagnostic imaging, laboratory tests, and cancer treatment has expanded greatly in the past two decades (Table 1-4). Although the amount of radioisotope administered per test has generally decreased, the total number of tests now exceeds 10 million per year (SNM, 1988). If radioimmunoassays (which use radioactive iodine) are included, the number of procedures exceeds 100 million per year. These methods are essential to modern medicine as they provide otherwise unobtainable information about organ function and disease status. Furthermore, they can often reduce or even eliminate the need for exploratory surgery. Relatively large amounts of isotope (up to about 20 mCi or 74 MBq) are used in clinical procedures and many of these radionuclides are potent gamma-ray emitters. Fortunately, the clinically important radionuclides such as ^{99m}Tc and ^{67}Ga are short-lived, with half-lives of less than 10 days. Many institutions also use in-house accelerators to produce ion beams for patient therapy that generate almost no waste except for some very short-lived (<1 to 2 hours) materials.

Government Generators

Nonmilitary government sources of LLRW include a diversity of facilities ranging from some Veteran's Administration hospitals to national, state, and local regulatory agencies. Consequently, the waste stream from these sources can mimic utility, institutional, or clinical

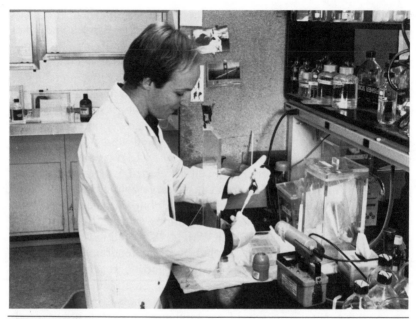

FIGURE 1-11. (*a*) Biomedical researcher using microcurie (kilobecquerel) amounts of ^{32}P to prepare a DNA probe to identify genes from the organism which causes sleeping sickness (trypanosomiasis); (*b; next page*) analyzing a DNA autoradiogram made with the ^{32}P probe screened against bacterial clones that contain various segments of DNA from *Trypanosoma brucei*.

TABLE 1-4. Principal Clinically Administered Radioisotopes[a]

Radionuclide	Principal Uses	Half-life	Typical Dose (mCi)[b]	Number of Procedures[c]	Total Curies[b]	% of Total
^{99m}Tc	Organ imaging	6.0 hours	4–25	8,040,000	116,580	96
^{201}Tl	Myocardial & parathyroid imaging	74.0 hours	2	960,000	1,920	2
^{67}Ga	Tumor/infection diagnosis	78.1 hours	5	600,000	3,000	2
^{131}I	Thyroid imaging	8.1 days	0.1	960,000	96	<0.1
Total				10,560,000	121,596	100

[a] SNM, 1988
[b] 1 Ci = 3.7×10^{10} Bq
[c] Annual number of procedures in the United States

sources. Companies performing government contract work usually do not appear on waste manifests as government sources; LLRW from many federal remedial action cleanup programs is also shipped to commercial disposal sites.

Greater than Class C Waste

Some radioactive wastes have activities that exceed the NRC's LLRW classification system. Materials exceeding the concentration limits established for Class C waste, known, not surprisingly, as "greater than Class C" (GTCC) waste, are specifically prohibited from commercial LLRW disposal sites (10 CFR Part 61). Through 1985, the last year for which data are available, about 2.4×10^6 Ci (8.88×10^{16} Bq) had accumulated in 130 m^3, giving a concentration of 18,500 Ci/m^3 (684 TBq/m^3) (Knecht, 1988). This is similar in magnitude to HLW (Figure 1-1). More than 95% of the activity and 32% of the volume of commercial GTCC waste is from the nuclear fuel cycle and is primarily due to the decontamination and decomissioning activities at several nuclear power plants. However, thousands of lower-activity nonutility generators make up the remainder of the GTCC inventory. These nonutility generators include manufacturers of sealed sources used in moisture gauges, well logging, and nondestructive testing of welds and piping; manufacturers of sealed sources used in industries, universities, and hospitals for sterilizing equipment and performing radiation therapy for cancer treatment; and companies involved in the manufacture of radiochemicals (McCabe, 1989) (Figure 1-12). Outside of Chernobyl and the bombings in World War II, accidents involving the inappropriate disposal of sealed sources with scrap metal, such as the incidents in Brazil and Mexico, have produced the most severe radiation exposures to members of the public (IAEA, 1988c).

GTCC waste is expected to approach an annual generation rate of about 170 m^3 by the year 2000 as larger numbers of nuclear reactors go off-line and are decommissioned. Estimates for the activity represented by individual radionuclides indicate that 54% of the activity in GTCC is from ^{3}H, ^{60}Co, and nuclides with half-lives shorter than 5 years. Since these nuclides may be disposed of without concentration limits as Class B LLRW, their segregation and removal, where possible, from the GTCC waste stream could significantly reduce the amount of radioactivity awaiting disposal.

Not classifiable as HLW, GTCC wastes can be disposed of only at specially licensed facilities that, to date, have not been developed. The DOE was assigned this responsibility by the Low-Level Radioactive Waste Amendments Act of 1985 and is required to develop capacity for GTCC disposal by 1989. The NRC has proposed that these wastes be placed in deep geologic repositories (10 CFR Part 61). A recent Office of Technology Assessment report (1988) concurred with this proposal and further suggested that, since the GTCC volume is very small, it be disposed of along with HLW. This suggestion has been acted upon, as the NRC released a final rule (*Federal Register*, 1989) amending regulations to require disposal of GTCC wastes in a

FIGURE 1-12. Because one-quarter of the world's food production is lost after harvesting, various foods are exposed to radioactive cobalt or cesium sources in order to control contamination and decay and thereby extend their safe consumption and shelf life. *(Courtesy of the International Atomic Energy Agency.)*

deep geologic repository. The NRC chose this option instead of revising its definition of HLW so as not to preclude alternative methods for GTCC disposal in the future.

Environmental Releases from Nuclear Power Plants

Routine and emergency releases of radionuclides from nuclear power plants are anticipated and permitted by both the NRC and EPA (10 CFR Part 20; 40 CFR Part 190). These releases are not, however, considered waste; instead, they are necessary by-products of the electricity generated by nuclear power. During ^{235}U decay, radioactive fission products are produced. Neutrons can also be captured by initially nonradioactive elements and become "activated" to produce radioactive forms of iron, nickel, chromium, etc. Although there is a very large inventory of radioactivity in a reactor core, multiple barriers and treatment systems reduce releases to low levels. These barriers begin with the fuel rod itself and its coating (known as cladding), the primary and secondary coolant systems, and finally the external containment vessel, which is designed to prevent releases that escape the primary cooling systems. Treatment includes the use of ion-exchange columns and particulate filters to clean up reactor cooling water and scrubbers and charcoal filters for removing some of the volatile isotopes from reactor gases.

The actual amounts of radioactive material released from a power plant vary greatly by the type of reactor design [pressurized (PWR) or boiling water (BWR) reactor], plant age, and physical maintenance of the reactor and steam-generating equipment. However, calculations for model reactors of both the PWR and BWR designs indicate that an average of 8 Ci (0.296 TBq) of ^{14}C (as $^{14}CO_2$), 25 Ci (0.925 TBq) of ^{41}Ar (argon gas), and about 700 Ci (25.9 TBq) of ^3H (as tritiated water, ^3HOH) are released annually from a typical 3500 MWt nuclear power plant (NCRP, 1987). Other radionuclides are released, including the long-lived species ^{129}I, but most have half-lives shorter than 24 hours and do not constitute a public health problem. Nevertheless, when these release rates are multiplied by the approximately 100 operating commercial nuclear power plants in the United States, the aggregate of released radioactivity is enormous and overshadows the entire inventory of LLRW disposed of by institutional and nonmilitary government sources. As an aside, the NRC's Atomic Safety and Licensing Board has given clearance for the evaporation of tritium-contaminated water at the Three Mile Island plant (Rosenstein, 1989). Over 1000 Ci (3.7×10^{13} Bq) of tritium is contained in this water.

Wastes below Regulatory Concern (BRC)

Congress asked the NRC, through the Amendments Act of 1985, to promulgate below-regulatory-concern (BRC) standards for the disposal of slightly contaminated wastes by means other than a commercial radioactive waste disposal facility. In August 1986, the NRC published guidelines, listed below from 51 *Federal Register* 30839, for petitioning to exempt specific waste streams from regulated disposal (10 CFR Part 2 Appendix B), providing a mechanism for petitioning for a generic ruling.

- Insignificant impact on the quality of human environment
- Maximum dose to any individual up to a few mrem (1 mrem = 0.01 mSv) a year
- Small collective dose to the critical and general populations

- No radiologically significant accidents or equipment malfunctions occur after loss of institutional controls
- A significant reduction in societal costs for LLRW disposal
- Proposed pretreatment and disposal options compatible with waste form
- Applicable to a significant portion of LLRW
- Radiological properties of the waste stream characterized on a national basis
- Data based on actual waste statistics
- Negligible potential for recycling of waste
- Licensee's program allows for licensing and inspection
- Off-site treatment and disposal facilities require no monitoring for radiation protection purposes
- BRC and uncontaminated wastes managed and assessed by equivalent procedures
- No regulatory or legal obstacles prohibit the proposed treatment or disposal methods

In December 1988, the NRC published a "Draft advance notice of the development of a Commission policy on exemption from regulatory control practices whose public health and safety impacts are below regulatory concern" (*Federal Register*, 1988). This notice suggested a maximum individual dose of 10 mrem/yr (0.1 mSv/yr) and 100 mrem/yr (1 mSv/yr) for estimated cumulative exposure to all exposed individuals from multiple BRC sources. At the same time, the EPA proposed a 4 mrem/yr (0.04 mSv/yr) maximum individual dose with a small collective dose (Holcomb et al., 1989). The Health Physics Society (Taylor, 1989a) agreed with the NRC's individual dose proposal but indicated that there should not be a limiting collective dose. The NRC intends to make this rule a "matter of compatibility," that is, that all Agreement States must accept the NRC's BRC ruling.

Two petitions have been prepared for the NRC that have a generic character, one for biomedical institutions (*Federal Register*, 1988) and another for the nuclear power industry by the Electric Power Research Institute (EPRI), which is pending submission as of press time. The first petition proposes on-site incineration of solid biomedical waste (paper, glass, plastics) containing a maximum of 1 Ci (37 GBq) of ^{3}H and 100 mCi (3.7 GBq) of ^{14}C per year. The ash would be disposed of in a sanitary landfill. The IMPACTS-BRC (Fortom and Goode, 1986) program used to evaluate the impact from this scenario assumes that 25% of ^{14}C and 10% of ^{3}H remain in the ash even though it is generally recognized that 100% of these radionuclides would be released from the incinerator stack. Therefore, no radiation impact is expected from leachate accumulation, groundwater transport, intruder, or transportation accidents that could occur at a landfill. The only impact would be to the off-site population and waste handlers at the institution. These exposures are 0.55 mrem/yr (0.0055 mSv/yr) maximum to an individual who spends 8 hours a day in the middle of the plume 100 m from the stack and 0.0034 mrem/yr (0.00034 mSv/yr) maximum worker exposure. To put these numbers into perspective, consider that one transcontinental (United States) airline flight delivers a radiation dose of about 2 mrem (0.02 mSv) from cosmic rays in the upper atmosphere (Eisenbud, 1988). The second petition proposes that nuclear power plant dry active waste (DAW), spent oil, soils and sludges, ion-exchange resins, grit blast material, and sewage and pond sludges be disposed of in sanitary landfills. The petition limits the concentration of gamma-emitting nuclides in the BRC waste and the total quantity in every bag of DAW as well as the total quantity disposed of annually from each reactor. The major radionuclides in these wastes, in decreasing prevalence, are ^{60}Co, ^{55}Fe, ^{63}Ni, ^{137}Cs, ^{54}Mn, ^{134}Cs, and ^{106}Ru. The maximum yearly exposure from this waste stream is received by the transportation workers in any disposal option. The dose is dominated by the external gamma irradiation of transportation workers from ^{60}Co. All other workers receive less than 4 mrem/yr (0.04 mSv/yr). The proposed dose limit of 15 mrem/yr (0.15 mSv/yr) to the transportation worker places an annual reactor limit of 90 μCi (3.33 MBq) for gamma radionuclides, except for spent oil, which has an annual limit of 4.4 μCi (162.8 kBq). The individual and population doses from this proposal are lower than that from regulated disposal at a commercial radioactive disposal facility because of the longer travel distances involved in transporting waste to existing LLRW facilities (Robinson et al., 1989).

Both petitions will result in dramatic savings in disposal space requirements. The EPRI proposal would result in a 27% reduction in LLRW volume from nuclear power plants and the biomedical proposal would eliminate over 90% of the institutional wastes (Party et al., 1989) that are presently shipped for commercial disposal. Reducing the BRC dose limit from 15 mrem/yr (0.15 mSv/yr) to 4 mrem/yr (0.04 mSv/yr) to meet the EPA proposal would result in expenditures of about $70 million per health effect avoided from these practices (Robinson et al., 1989). According to the EPA, adoption of their recommended BRC level would reduce the total LLRW volume from all generators by 35% (Holcomb et al., 1989).

Mixed Waste

The EPA, under the Resource Conservation and Recovery Act (RCRA), promulgated a listing of hazardous materials banned from land disposal based on the characteristics of ignitability, corrosivity, reactivity, and toxicity (40 CFR Part 261). LLRW with these additional characteristics is known as "mixed waste." It is generated by virtually all types of users of radionuclides and consists of contaminated organic solvents, oils, lead shielding, and chromate solutions (Bowerman et al., 1986). The NCR's deregulation of liquid scintillation wastes in 1981 (10 CFR Part 20.306) eliminated a major component of mixed wastes. Although the remaining mixed wastes account for only a small fraction (2–5%) of all LLRW, their disposal at existing sites has been banned.

Mixed wastes present land burial problems because the nonradioactive components are hazardous and may promote the mobility of radionuclides. They also present regulatory authority problems since these wastes are under the authority of the EPA, NRC, and different state agencies under different statutes (Leventhal and Kharkar, 1985). It is now the responsibility of generators to identify and properly manage mixed wastes. At the present time, disposal options do not exist for mixed wastes nor may they be legally stored by the generator for more than 90 days unless the facility has an RCRA Part B permit, which is difficult to obtain. The EPA and NRC have failed to resolve this problem.

Exempt Wastes

Some radiation sources are effectively exempt from 10 CFR Part 61 disposal requirements because they can be obtained without a radioactive materials possession license. These are certain broadly distributed medical supplies such as radioimmunoassay kits and manufactured goods such as ionizing smoke detectors, tritium-powered emergency exit signs, and instruments and jewelry with radioluminescent dials (NCRP, 1977; Eisenbud, 1987; NSC, 1987). Loopholes in the regulations as well as inadequate enforcement of existing laws mean that many of these products end up in the ordinary trash. Manufacturers of such products must, however, comply with regulations found under 10 CFR Part 31. These products frequently contain warning labels and disposal instructions that are usually ignored by consumers. Even if consumers sought proper disposal, no such mechanism exists. These products may contain substantial amounts of radioactivity (see Table 1-5) and would be regulated as radioactive waste if they were generated for other purposes, but are often handled and disposed of at municipal landfills and incinerators without regard to their radioactivity.

Excreta from patients receiving radiopharmaceuticals for imaging scans and treatment, regardless of the level of contamination, are also exempt from disposal regulations. Although relatively high doses can be received by hospital personnel and family members in close proximity to these patients (Benedetto et al., 1989), the short half-life of most radiopharmaceuticals and their disposal by sewer release should minimize any possible health hazards off-site.

Unregulated Wastes

The disposal of naturally occurring (e.g., ^{226}Ra) and accelerator-produced radioactive materials (NARM) is not regulated by any federal agency, and state regulations are inconsistent or nonexistent, although some generators of NARM do ship these wastes for disposal. Texas

TABLE 1-5. Examples of Materials Exempt from Licensing Requirements under 10 CFR Part 31

Product[a]	Permissible Activity[b] ($\leq$)
Static-elimination devices	500 μCi ^{210}Po
Ion-generating tubes	500 μCi ^{210}Po or 50 mCi ^{3}H
Luminous devices in aircraft	10 Ci ^{3}H or 300 mCi ^{147}Pm
Calibration sources	5 μCi ^{241}Am
Ice-detection devices	50 μCi ^{90}Sr
Prepackaged in vitro/clinical testing kits	10 μCi ^{125}I/test 10 μCi ^{131}I/test 10 μCi ^{14}C 50 μCi ^{3}H 20 μCi ^{59}Fe 10 μCi ^{55}Fe

[a] The use of thorium in gas mantles, vacuum tubes, welding rods, incandescent lamps, some kinds of photographic films, and finished optical lenses is also not regulated. Naturally occurring radioactive materials (NORM) present in geologic specimens, petroleum drilling wastes, and rare earth minerals processing wastes (with the exception of uranium and thorium) are also not regulated.

[b] 1 Ci = 3.7 × 10^{10} Bq

is preparing to permit a permanent storage facility for naturally occurring radioactive materials (Helminski, 1989) and the inconsistency can be dramatic. For example, ^{109}Cd can be produced by both reactors and accelerators; the former is regulated, the latter is not. Although most accelerators are of too low an energy to create activation products, small amounts of very short-lived wastes are produced. Medical cyclotrons produce more radioactive wastes than simple accelerators, but the half-lives of their wastes are generally less than 2 hours. Approximately 100 cyclotrons produce radioactive gases used in conjunction with positron emission tomography (PET) scanners at major medical facilities across the country.

The EPA has proposed regulations (40 CFR Part 764) for NARM under the Toxic Substances Control Act. This proposal would subject wastes with more than 2 nCi/g (74 Bq) to strict disposal requirements similar to those for LLRW. Exemptions will be made for mining wastes and certain consumer products (Gruhlke et al., 1989).

Radionuclides in LLRW

Half-life and mobility have important implications for management policies and the design criteria of any radioactive materials disposal site. The principal radionuclides, their sources, and their approximate abundance in LLRW appear in Table 1-6. Regardless of source, most of the activity in LLRW is from nuclides with half-lives shorter than 100 years. In contrast to the other generators, about 70% of the activity of institutional LLRW is contaminated with nuclides that have half-lives of less than 90 days; clinically administered nuclides have even shorter half-lives. The predominant long-lived nuclide in all non-fuel-cycle wastes is ^{3}H, whereas in the fuel cycle the predominant long-lived nuclides are ^{60}Co, ^{134}Cs, and ^{137}Cs. Depending on the solubility of nuclides in air and water or that of the compounds that they compose, nuclides may migrate through the environment and be absorbed by humans directly or indirectly through our food and biogeochemical cycles. Dose calculations show that ^{3}H is the predominant dose-producing radionuclide over the short-term water pathway, while ^{14}C, ^{129}I, and ^{99}Tc predominate over the long-term pathway. Also, ^{14}C is the predominant dose-producing radionuclide along the airborne pathway at all times (Matuszek, 1988). The low molecular weight, reactivity, and ubiquity of carbon and hydrogen in living organisms make their radioisotopes particularly hazardous if taken internally.

TABLE 1-6. Typical Radionuclides Present in LLRW by Generator Category[a]

Radionuclide	Half-life	Activity (%)				
		Fuel Cycle	Industry	Institutional	Clinical	Government
<90 day half-life:						
^{99m}Tc	6.0 hours	...	...	8	96	...
^{99}Mo	66.7	...	...	3		...
^{201}Tl	74.0	...	...		1-2	...
^{67}Ga	78.1	...	...	3	1-2	...
^{131}I	8.1 days	...	...	1	0.2	...
^{32}P[b]	14.3	...	...	22	...	7
^{86}Rb	18.7	...	...	<1	...	7
^{125}I	60.2	...	...	9	...	7
^{192}Ir	74.2	...	7	<1	...	7
^{35}S[b]	88.0	...	<1	22	...	7
>90 days and <5 years half-life:						
^{134}Cs	2.1 years	18	<1	...	...	...
^{22}Na[b]	2.6	...	...	<1	...	7
>5 years and <100 years half-life						
^{60}Co	5.3 years	16	1	<1	...	...
^{3}H[b]	12.3	2	88	29	...	55
^{137}Cs	30.0	36	<1	<1	...	...
>100 years half-life						
^{14}C[b]	5,730 years	<1	<1	1	...	10
^{99}Tc	210,000	<1		2	...	...
Uranium and TRU	>10^5	...	3	<1	...	...
Mixed fission[b,c]	<10^6	25	...	...	...	...
Total		98	99	100	100	100

[a] NCRP, 1983; New York State Energy Office, 1984; Eisenbud, 1987; NLLRWP, 1988
[b] Naturally occurring isotope
[c] Mixed fission products include radioactive xenon, krypton, bromine, iodine, tellurium, ruthenium, strontium, and barium. With the exceptions of ^{129}I, ^{90}Sr, and ^{85}Kr, all have half-lives shorter than one year.

How Much LLRW Is Produced?

A breakdown of the LLRW received by the disposal sites in 1987 appears in Table 1-7 and shows that 97% of the volume of LLRW was Class A, with Class B and C wastes contributing little to the total. Conversely, the total activity is highest in Class C. Concentration spans three orders of magnitude, with the Class A waste averaging 0.50 Ci/m^3 (1.85 $\times$ 10^{10} Bq/m^3). Further increases in radioactive concentration should be anticipated as volume-reduction methods such as incineration and supercompaction become more commonplace.

The annual activity and volume of LLRW received for commercial disposal from 1979 to 1988 (Figure 1-13) show a rate of waste generation of about 60,000 to 80,000 m^3/yr, although a trend toward lower volume is becoming apparent. Concentrations of radioactivity generally increased over this period. We suspect this increase is mainly due to financial incentives to reduce waste volumes (primarily by solid waste compaction). Significant deviations in volume occurred in 1980 and 1986. In 1979 the three operating disposal sites were temporarily closed and caused LLRW inventories awaiting shipment for disposal to accumulate. This volume was carried into 1980 and is a likely explanation for this peak. Conversely, the marked decline in volume and activity in 1986 reflects the vigorous volume-reduction and waste-minimization measures instituted in that year at many nuclear power plants to comply with the Amendments Act. Given that the nuclear fuel cycle accounts for more than 50% of all LLRW (Table 1-8), the annual volume of LLRW will be heavily influenced by power plant refueling and operational cycles. Institutional wastes have the lowest activity and volume of all LLRW. To place all of these numbers into perspective, consider that they are a small fraction of the global ra-

TABLE 1-7. LLRW Shipped to Commercial Disposal Sites in 1987, by Class[a]

Class	Activity Ci[b]	%	Volume m^3	%	Concentration (Ci/m^3)[b]
A	26,068	9.6	50,877	97.4	0.51
B	66,981	24.9	1,108	2.1	60.45
C	176,501	65.5	246	0.5	717.48
Total	269,550	100.0	52,231	100.0	5.16

[a] NLLRWMP, 1988
[b] 1 Ci = 3.7 $\times$ 10^{10} Bq

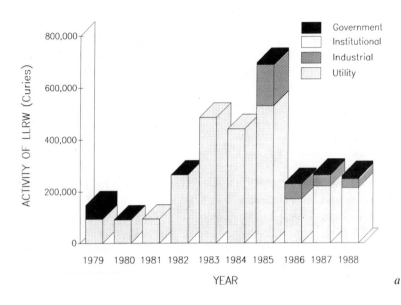

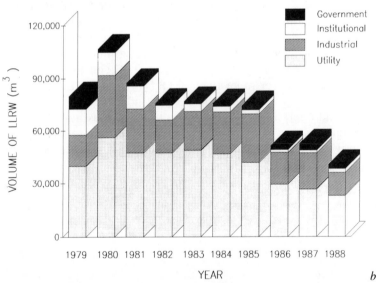

FIGURE 1-13. (*a*) Activity and (*b*) volume of low-level radioactive wastes shipped for commercial disposal from 1979 to 1988. *(Data courtesy of NLLRWMP, 1980–1989.)*

TABLE 1-8. Average Annual Activity and Volume of Commercial LLRW Received at Disposal Sites for 1985–1988, by Generator Group[a]

Generator Class	Number of Generators	Activity		Volume		Concentration Ci/m^{3b}
		Ci[b]	%	m^3	%	
Utility	113	283,155	77.7	30,125	55.3	9.40
Industrial	716	73,125	20.1	19,832	36.4	3.69
Institutional	508	867	0.2	1,823	3.3	0.48
Government	116	7,408	2.0	2,737	5.0	2.71
Total	1453	364,555	100.0	40,461	100.0	6.69

[a] NLLRWMP, 1986–1989
[b] 1 Ci = 3.7×10^{10} Bq

dioactive inventory (see Chapter 6) and military and power plant HLW (Figure 1-1).

The nine leading LLRW-producing states in 1987 were, in decreasing order by volume, Illinois, Pennsylvania, South Carolina, New York, California, Tennessee, Oklahoma, Oregon, and North Carolina. Together these states accounted for about 50% of the activity and 60% of the volume of all LLRW shipped for disposal in 1988 (Figure 1-14). Two basic patterns of waste generation are evident with either utility or industrial sources dominating the state profile. In four of these states, utility sources predominate, but for the remainder, the leading source is industrial. In 1987 Tennessee erroneously became the nation's largest volume generator when the national LLRW survey (NLLRWMP, 1988) improperly credited Tennessee for wastes processed in the state by treatment facilities that began commercial operation in that year. Although utilities and industry produce the largest amounts of LLRW, and although some regions produce more than others, the entire country shares many of the benefits derived from the use of radioactive materials manufactured or otherwise used in these states. For example, in New York State one manufacturer (Cintichem) accounts for nearly 60% of all industrial LLRW produced in the state and has a significant impact on the total volume of waste reported for the state (NYSERDA, 1989), but it is also the nation's largest single source of ^{99m}Tc used in medicine.

A particularly serious problem for the entire LLRW issue is the paucity of uniform, high-quality data on the content and source of wastes. Survey inconsistencies in many of the early state-by-state

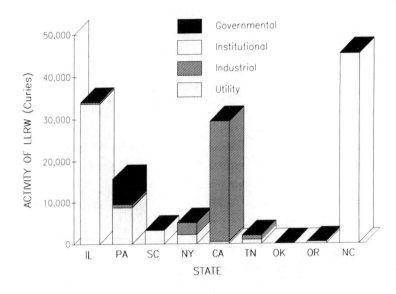

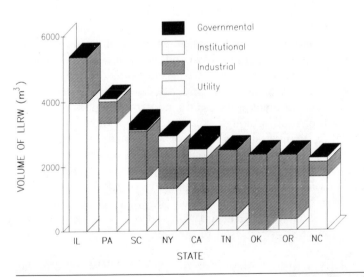

FIGURE 1-14. (*a*) Activity and (*b*) volume of low-level radioactive wastes shipped for commercial disposal by the primary waste-generating states in 1988. (*Data courtesy of NLLRWMP, 1989.*)

assessments conducted by DOE led to the incomplete characterization of non-fuel-cycle wastes. A uniform national manifest system has never been established for radioactive wastes, although the different manifests used must comply with the regulations under 10 CFR Part 61. Since as many as four or five government agencies and private contractors may be involved in preparing national LLRW disposal data, their multiple reentry and processing leads to additional sources of error. The companies that operate the commercial disposal facilities compile information from what is provided on shipping manifests, but no detailed national database exists. These companies are unresponsive to requests for data and make it available only at a high fee, thereby placing an important source of historical data out of the reach of many analysts. Even when available, manifest data are usually inaccurate since most individual generators do not have their own internal reporting systems. This is especially problematic for institutions and brokers that collect a wide variety of wastes for which no single treatment method is adequate. Since more than 95% of LLRW is Class A, the low energy and activity of most of this waste means that accurate external surveys are difficult to perform, providing another incentive for good recordkeeping practices that are essential to the proper management of LLRW. Without actual data from LLRW generators, it has proven difficult to predict accurately the number and size of new disposal sites and also to model the impacts from different waste disposal scenarios. Furthermore, inadequate data on LLRW is one reason why the NRC has failed to codify the deregulation of very LLRW (*Federal Register,* 1986).

Prior to the late 1970s, little effort was placed on characterizing LLRW, and most regulations governing its disposal were written with the expectation of shallow land burial. Recent interest in highly engineered structures instead of shallow land burial have focused on retrievability of the wastes. While the commissioner of the NRC has always held the power to grant license exemptions for alternatives to the generic disposal methods, petitioning for exemption is a difficult and lengthy process (Higginbotham et al., 1986). Agreement state status, whereby a state or local government agency may assume the NRC's responsibility for regulating nuclear materials with stringency equal to or greater than the NRC, further complicates the issue by expanding the number of pertinent jurisdictions.

The NRC's classification system for LLRW has several shortcomings. Since the majority of waste is Class A, and much of this is of very low activity, subdivision(s) of the Class A limits could relax the design stringency at disposal facilities. For example, very short-lived materials (<90 to 180 days) should not be accepted by disposal facilities at all and should instead be decayed on-site or at a specially designated facility. Waste containing isotopes with half-lives equal to or shorter than that of ^{3}H (12.3 years) could be separately packaged and disposed of by long-term storage alone. These wastes would decay to low levels of radioactivity within 100 years and to trivial levels by 200 years. High surface exposures could be problematic, however, if isotopes such as ^{22}Na, ^{60}Co, and ^{134}Cs are present, and would require shielding. Class A wastes containing long-lived radionuclides, even in low activity, should be subjected to the more stringent stability requirements usually reserved for Class B and C waste.

In the face of such regulatory inconsistencies as the disposal exemptions for patient excreta, radioactive consumer products, and the very large environmental releases permitted by nuclear power plants, petitioning for case-by-case rulings for LLRW is unnecessarily burdensome on generators of low-activity wastes. We believe that both the public's health and pocketbook would be better protected if rulings were instead initiated by the NRC to deal with disposal issues affecting multiple generators or entire groups of generators. These should be based solely on the characteristics of the waste streams, provided that the generators are willing and able to supply the data necessary to conduct these studies.

If nothing else, our current politically motivated, decentralized approach to LLRW management may ultimately yield more useful characteriziations of regional and local waste streams. Perhaps one day this information will be unified under federal leadership to better manage a problem that is clearly national in scope.

References

Benedetto, A. R., T. W. Dziuk, and M. L. Nusynowitz. 1989. Population exposure from nuclear medicine procedures: measurement data. *Health Physics* 57(5):725–731.

Booth, W. 1987. Postmortem on Three Mile Island. *Science* 238:1342–1345.

Bowerman, B. S., R. E. Davis, and B. Siskind. 1986. *Document Review Regarding Hazardous Chemical Characteristics of Low-Level Waste.* NUREG/CR-4433. Upton, N.Y.: Brookhaven National Laboratory.

Caldwell, J. A., M. Kyllo, M. Matthews, and J. D'Antonio. 1989. DOE disposal cell design and surveillance and maintenance programs. *Waste Management '89* 2:29–35.

Cohen, B. L. 1986. A generic probabilistic risk analysis for a high-level waste repository. *Health Physics* 51(4):519–528.

Council on Scientific Affairs (CSA). 1989. Medical perspective on nuclear power. *Journal of the American Medical Association* 262(19):2724–2729.

Department of Energy (DOE). 1988. *Data Base for 1988: Spent Fuel and Radioactive Waste Inventories, Projections, and Characteristics.* DOE/RW-0006, Rev. 4. Washington, D.C.: DOE.

Eisenbud, Merril. 1987. *Environmental Radioactivity.* Orlando, Fla.: Academic Press.

Eisenbud, Merril. 1988. Low-level radioactive waste repositories: A risk assessment. *Annual Conference of the North Carolina Academy of Sciences,* 26 March 1988, Charlotte, N.C.

Federal Register. 1986. 51:30839.

Federal Register. 1988. 53:43896.

Federal Register. 1989. 54:22578.

Fortom, J. M., and D. J. Goode. 1986. *De Minimis Waste Impacts Analysis Methodology: IMPACTS-BRC User's Guide and Methodology for Radioactive Wastes Below Regulatory Concern.* NUREG/CR-3585. Washington, D.C.: NRC.

Gertz, C. P. 1989. Yucca Mountain, Nevada: Is it a safe place for isolation of high-level radioactive waste? *Waste Management '89* 1:9–12.

Gruhlke, J. M., F. L. Galpin, W. F. Holcomb, and M. S. Bandrowski. 1989. U.S. EPA's proposed environmental standards for the management and land disposal of LLW and NARM waste. *Waste Management '89* 2:273–276.

Halverson, T. W. 1989. TRUPACT II—The story of success. *Waste Management '89* 1:235–238.

Helminski, E. L. 1989. In Texas. *The Radioactive Exchange* 8(16):7.

Higginbotham, L. B., K. S. Drogonette, and C. L. Pittiglio, Jr. 1986. *Licensing of Alternative Methods of Disposal of Low-Level Radioactive Waste.* NUREG-1241. Washington, D.C.: NRC.

Holcomb, W. F., J. M. Gruhlke, and F. L. Galpin. 1989. US EPA's proposed standard for BRC criteria. *Waste Management '89* 2:361–364.

Houk, V. N. 1989. Epidemiology in risk evaluation. *Bulletin of the New York Academy of Medicine* 65(4):461–466.

International Atomic Energy Agency (IAEA). 1987. *Safe Management of Wastes from the Mining and Milling of Uranium and Thorium Ores.* Safety Series No. 85. Vienna, Austria: IAEA.

International Atomic Energy Agency (IAEA). 1988a. *Nuclear Power and Fuel Cycle: Status and Trends.* Vienna, Austria: IAEA.

International Atomic Energy Agency (IAEA). 1988b. *Nuclear Research Reactors in the World.* Reference Data Series No. 3. Vienna, Austria: IAEA.

International Atomic Energy Agency (IAEA). 1988c. *Nuclear Safety Review for 1987.* Vienna, Austria: IAEA.

Jones, K. L. 1989. RCRA Part A and Part B compliance. Paper read at 34th Health Physics Annual Meeting, 25–29 June 1989, Albuquerque, New Mexico.

Kempf, C. R. 1985. Alternatives for packaging C-14 waste: C-14 generator survey summary. A-3172. Upton, N.Y.: Brookhaven National Laboratory.

Klein, D. E. 1989. Status and projected activities of the monitored retrievable storage review commission. *Waste Management '89* 1:13–20.

Knecht, M. A. 1988. Greater-than-Class-C low-level radioactive waste management concepts. *Waste Management '88* 1:637–641.

Ko, S. 1988. Analysis of the reduction in waste volumes received for disposal at the low-level radioactive waste site in the state of Washington. *Waste Management '88* 1:629–635.

Leventhal, L., and D. P. Kharkar. 1985. Laboratory experience in the analysis of orphan waste. Paper read at 28th Annual Oak Ridge National Laboratory Department of Energy Conference, Knoxville, Tennessee.

Matuszek, J. M. 1988. Safer than sleeping with your spouse—the West Valley experience. In *Low-Level Radioactive Waste Regulation,* ed. Michael E. Burns, Chelsea, Michigan: Lewis Publishers, pp. 261–277.

MacKenzie, D. R. 1988a. Characterization of low-level waste from the industrial sector, and near-term projection of waste volumes and type. *Waste Management '88* 1:589–594.

MacKenzie, D. R. 1988b. *Characterization of Low-Level Waste from the Industrial Sector, and Near-Term Projection of Waste Volumes and Type.* BNL/NUREG-40927. Upton, N.Y.: Brookhaven National Laboratory.

McBride, J. P., R. E. Moore, J. P. Witherspoon, and R. E. Blanco. 1978. Radiological impact of airborne effluents of coal and nuclear plants. *Science* 202:1045–1050.

McCabe, G. H. 1989. A management approach for greater-than-Class-C LLRW. *Waste Management '89* 2:5–9.

National Academy of Science (NAS). 1957. *Report to the U.S. Atomic Energy Commission: Disposal of Radioactive Waste on Land.* Washington, D.C.: NAS.

National Council on Radiation Protection and Measurements (NCRP). 1977. *Radiation Exposure from Consumer Products and Miscellaneous Sources.* NCRP Report No. 56. Washingon, D.C.: NCRP.

National Council on Radiation Protection and Measurements (NCRP). 1983. *Environmental Radioactivity.* NCRP Proceedings No. 5. Bethesda, Md.: NCRP.

National Council on Radiation Protection and Measurements (NCRP). 1987. *Public Radiation Exposure from Nuclear Power Generation in the United States.* NCRP Report No. 92. Bethesda, Md.: NCRP.

National Low-Level Radioactive Waste Management Program (NLLRWMP). 1980. *The 1979 State-by-State Assessment of Low-Level Radioactive Wastes Received at Commercial Disposal Sites.* NUS-3340. Washington, D.C.: DOE.

National Low-Level Radioactive Waste Management Program (NLLRWMP). 1982a. *The 1980 State-by-State Assessment of Low-Level Radioactive Wastes Received at Commercial Disposal Sites.* LLWMP-11T. Washington, D.C.: DOE.

National Low-Level Radioactive Waste Management Program (NLLRWMP). 1982b. *The 1981 State-by-State Assessment of Low-Level Radioactive Wastes Received at Commercial Disposal Sites.* DOE/LLW-15T. Washington, D.C.: DOE.

National Low-Level Radioactive Waste Management Program (NLLRWMP). 1983. *The 1982 State-by-State Assessment of Low-Level Radioactive Wastes Received at Commercial Disposal Sites.* DOE/LLW-27T. Washington, D.C.: DOE.

National Low-Level Radioactive Waste Management Program (NLLRWMP). 1984. *The 1983 State-by-State Assessment of Low-Level Radioactive Wastes Received at Commercial Disposal Sites.* DOE/LLW-39T. Washington, D.C.: DOE.

National Low-Level Radioactive Waste Management Program (NLLRWMP). 1985. *The 1984 State-by-State Assessment of Low-Level Radioactive Wastes Received at Commercial Disposal Sites.* DOE/LLW-50T. Washington, D.C.: DOE.

National Low-Level Radioactive Waste Management Program (NLLRWMP). 1986. *The 1985 State-by-State Assessment of Low-Level Radioactive Wastes Received at Commercial Disposal Sites.* DOE/LLW-59T. Washington, D.C.: DOE.

National Low-Level Radioactive Waste Management Program (NLLRWMP). 1987a. *The 1986 State-by-State Assessment of Low-Level Radioactive Wastes Received at Commercial Disposal Sites.* DOE/LLW-66T. Washington, D.C.: DOE.

National Low-Level Radioactive Waste Management Program (NLLRWMP). 1987b. *LLRW Management in Medical and Biomedical Research Institutions.* DOE/LLW-13Th. Washington, D.C.: DOE.

National Low-Level Radioactive Waste Management Program (NLLRWMP). 1988. *The 1987 State-by-State Assessment of Low-Level Radioactive Wastes Received at Commercial Disposal Sites.* DOE/LLW-69T. Washington, D.C.: DOE.

National Low-Level Radioactive Waste Management Program (NLLRWMP). 1989. *The 1988 State-by-State Assessment of Low-Level Radioactive Wastes Received at Commercial Disposal Sites.* Washington, D.C.: DOE (in preparation).

National Safety Council (NSC). 1987. More problems with glow-in-the-dark exit signs. *Campus Safety Newsletter.* Chicago: NSC.

Neill, R. H., and L. Chaturvedi. 1989. Technical and programmatic evaluation of WIPP. *Waste Management '89* 1:253–259.

New York State Energy Office. 1984. *Low-Level Radioactive Waste Management Study, Main Report,* Vol. 2. New York: New York State Energy Office.

New York State Energy Research and Development Authority (NYSERDA). *1988 New York State Low-Level Radioactive Waste Status Report.* New York: NYSERDA.

Office of Technology Assessment (OTA). 1988. *An Evaluation of Options for*

Managing Greater-than-Class-C Low-Level Radioactive Waste. Washington, D.C.: OTA.

Party, E., and E. L. Gershey. 1989. Recommendations for radioactive waste reduction in biomedical/academic institutions. *Health Physics* 56(4):571–572.

Party, E., A. Wilkerson, and E. L. Gershey. 1989. Need for broad generic BRC rulings for biomedical low-level radioactive waste. *Radiation Protection Management* 6(6):45–51.

Robinson, P. J., J. N. Vance, and V. Rogers, 1989. Summary of EPRI BRC research program. *Waste Management '89* 2:379–382.

Roche-Farmer, L. 1980. *Study of Alternative Methods for the Management of Liquid Scintillation Counting Wastes*. NUREG-0656. Washington, D.C.: NRC.

Rosenstein, M. 1989. A musing column: Another phase. *The Health Physics Society's Newsletter,* Volume XVII, No. 3 (March).

Schurr, S. H., J. Darmstadter, H. Perry, W. Ramsay, M. Russell. 1979. *Energy in America's Future: The Choice Before Us*. Baltimore, Md.: Johns Hopkins University Press.

Society of Nuclear Medicine (SNM). 1988. *Nuclear Medicine Self-Study Program 1*. New York: SNM.

Taylor, L. S. 1989a. HPS comments on NRC policy statement. *The Health Physics Society's Newsletter,* Volume XVII, No. 3 (March).

Taylor, J. T. 1989b. Improved and safer nuclear power. *Science* 244:318–325.

United Nations Scientific Committee on the Effects of Atomic Radiation (UNSCEAR). 1977. *Sources and Effects of Ionizing Radiation*. New York: United Nations.

2 HISTORY OF COMMERCIAL DISPOSAL

Since the birth of the nuclear age in the 1940s, the United States has employed two different disposal methods for low-level radioactive waste (LLRW): the wastes either have been buried on land or dumped into the ocean. In 1960, the Atomic Energy Commission (AEC) stopped issuing permits for ocean disposal and began opening government land burial facilities, such as those at Oak Ridge, Tennessee, and Idaho Falls, Idaho. Less and less LLRW was sent for ocean disposal, and by 1962, approximately 95% of all LLRW was disposed of at shallow land burial facilities. (See Chapter 4 for a description of different disposal methods.) Although federally funded labs were accepting waste from private generators, by 1960 the rapid growth in the use of radioactive materials and consequent generation of LLRW by the private sector led the AEC to propose that commercial sites be developed.

Commercial disposal of LLRW started in 1962 at the Beatty, Nevada, site. Within nine years, five other commercial sites had opened. Only three of the original six sites remain in operation today. The current status, along with the general physical characteristics, of all six sites is shown in Table 2-1. All the sites employ shallow land burial (SLB). Initially, the sites were licensed and operated with limited regulation and poorly defined performance criteria (Godbee and Kibbey, 1983). Standards for the methods of disposal were not promulgated until 1981 (10 CFR Part 61). Segregation of transuranic (TRU) waste from LLRW was not required until 1970, nor were the packaging and shipping requirements for LLRW as rigorous as they are now. The commercial sites did not stop burying TRU wastes until 1979. TRU wastes alone account for approximately 30% of the total volume of waste buried at all of the commercial sites except Barnwell, which

TABLE 2-1. Status and Characteristics of Commercial LLRW Disposal Sites[a]

	Maxey Flats	West Valley	Sheffield	Barnwell	Beatty	Richland
Start-up and closure	1962–1977	1963–1975	1967–1978	1971–	1962–	1965–
Operator	US Ecology	Nuclear Fuel	US Ecology	Chem-Nuclear	US Ecology	US Ecology
Licensing authority	State	State	NRC	State & NRC	State	State & NRC
Total facility area	280 acres	3345 acres	320 acres	300 acres	80 acres[b]	1000 acres
Burial site area	25 acres	12 acres	20 acres	47 acres	47 acres	100 acres
Mean annual precipitation (mm)	1050	1040	900	1200	100	172
Surface material	Clay, siltstone	Till, gravel silty clay	Silt, sand	Sand, clay	sand gravel	Silt, sand gravel zones
Interstitial permeability	Low	Low	Low	Low	Moderate	Low
Bedrock material	Shale, silt-stone, sandstone	Shale, siltstone	Shale, limestone	Sedimentary sands	Clay, shale	Basaltic lavas
Depth to groundwater (m)	Unknown	31–38	6–15	10–20	100	Unknown
Depth to regional aquifer (m)	85	>60	>50	200	Unknown	110

[a] Dayal et al., 1986; NYDEC, 1987c
[b] Surrounded by a buffer zone of 400 acres

48

never accepted TRU wastes (Godbee and Kibbey, 1983). In essence, the sites were operated little differently than ordinary landfills.

The inventory of LLRW buried at the six sites represents approximately 30% of all LLRW buried in shallow land sites in the United States; the remainder, primarily military and defense wastes, is held in DOE sites (Carter and Stone, 1986) (Figure 1-1). The history of the commercial sites provides significant information on the management of commercial LLRW.

Closed Sites

Well-publicized problems associated with water management led to the closure of the Maxey Flats site. The West Valley commercial facility was voluntarily closed in 1975 by the site operator in response to political pressure following disclosure of problems with water accumulation in some of the trenches, but the NRC-licensed facility there continued to operate until 1986. The Sheffield facility closed when it was unable to acquire a permit for expansion from the NRC (Mallory, 1983).

Maxey Flats, Kentucky

During its operation from May, 1963, to the end of 1977, this 280-acre facility, run by US Ecology, Inc., accepted 140,000 m^3 of LLRW at its 25-acre burial area. By 1972, some of the trenches had become partially or completely filled with infiltrated water and a water management program was instituted, mostly by pumping water from the trenches and evaporating the collected water. Although this program reduced water levels, it was insufficient to remedy the problem entirely. In 1976, state health officials temporarily closed the site when they discovered that radioisotopes, mainly ^{3}H, were migrating through fractured sandstone from the trenches to other parts of the site (Miller and Bennett, 1985). Furthermore, emissions from the evaporation of leachate pumped out of the trenches appeared to have caused contamination of a local milk supply (NCRP, 1987). Although studies by the EPA showed that contamination of the groundwater and milk was well within the EPA standards for drinking water (4 mrem/yr), the state canceled US Ecology's lease in 1977 (Montgomery et al., 1977;

Eisenbud, 1987). After closure in 1978, Kentucky bought the rights to the site from US Ecology. Part of this 1978 agreement provided a blanket indemnity for US Ecology against present and future liabilities for the site (Helminski, 1988b). The responsibility for decommissioning and cleanup was given to the Kentucky Department for Natural Resources and Environmental Protection in 1979 (NYSDEC, 1987c; NYSDEC, 1987f). In 1981, temporary plastic covers were installed over a number of the trenches in an effort to stop the influx of rainwater, but they were not completely effective (Dayal et al., 1986). The site was placed on the EPA's Superfund list of cleanup sites in 1985, with a target date for stabilization in 1993. Since 1985, Maxey Flats has been used as a testing ground for cleanup operations (Health and Environment Network, 1988). In 1986, the EPA began collecting funds from former users, operators, and owners of the site for remedial investigations and procedures (Helminski, 1988b).

The major problems at the site resulted from the relatively low permeability of the shale and till at the site, which led to accumulation of water in the trenches. Drums and packages in a wet trench are subject to rapid corrosion, after which the waste undergoes leaching and microbial degradation. The state is still spending in excess of $1,000,000 a year on monitoring and remediation (NYSDEC, 1987c). Westinghouse presently manages the site and estimates the total cost of closing the site at $25 million. In an effort to defray the cost to the state, Kentucky is attempting to collect additional fees from the generators who have waste buried at Maxey Flats and has filed suit to have the 1978 agreement with US Ecology declared unconstitutional. US Ecology is countering this with a suit for recovery of all costs associated with the EPA-ordered remediation efforts as well as for enforcement of the rights and claims of the 1978 agreement (Helminski, 1988b).

The problems resulting from water accumulation were exacerbated by the poor waste management practices employed during the early years of operation. Not only were containers of liquid accepted, but also packages of solid waste were simply dumped at random into the trenches and covered with 1 to 3 m of compacted clay and crushed shale. In addition to being poorly emplaced, many packages were damaged during emplacement or during compaction of the clay trench caps, with the overall effect of slowly compressing the waste, thereby creating approximately 50% void space in the trenches (Dayal et al., 1986). The high void space has led to subsidence of the trenches and instability of the caps.

West Valley, New York

The Western New York Nuclear Services Center (WNYNSC), developed by New York State on 3345 acres of farmland acquired through eminent domain, includes the world's first commercial nuclear fuel reprocessing plant, a 7-acre waste burial ground for long-lived wastes from the reprocessing activities, a spent fuel and high-level radioactive waste (HLW) receiving and storage facility, a low-level liquid waste treatment plant, and the commercial LLRW burial facility (NYSDEC,1987c) (Figure 2-1). Most of the facilities are located within a 250-acre plot in the center of the WNYNSC. Between November 1963 and March 1975, WNYNSC, operated by Nuclear Fuel Services, Inc. (NFS) and licensed by the state, accepted 66,837 m^3 of LLRW containing approximately 736,000 Ci (2.72 × 10^{16} Bq) (Anderson, 1988). The LLRW site was licensed to handle three types of LLRW: by-product materials (^{3}H, ^{14}C, ^{60}Co, ^{125}I, ^{131}I, ^{137}Cs, ^{241}Am), source materials (^{232}Th, natural uranium), and special nuclear materials (^{235}U,

FIGURE 2-1. Aerial view of the West Valley Demonstration Project disposal area showing both the WVDP facilities and the now-closed state LLRW disposal area (the flat grassy area in the upper right-hand corner). *(Courtesy of the West Valley Demonstration Project.)*

^{238}Pu, ^{239}Pu) (Anderson, 1988). These wastes were buried in twelve parallel trenches and two special pits in an area which occupies only 15 acres. A separate area, licensed by the NRC, was operated for the disposal of low-level wastes from the reprocessing facility at the site.

The trenches are located in two areas. The northern area contains five trenches (10 m by 180 m and 6 m deep), a shallow concrete trench, and a series of holes for high-activity wastes. Two lagoons were excavated and used to store rainwater pumped out of the trenches; these lagoons are now filled with soil. Lessons learned from water accumulation in the northern area led to improvements in the construction of the seven trenches developed in the southern area from 1969 to 1975. A third lagoon was excavated near the southern area for pumping water from completed trenches (Anderson, 1988).

The possibility of water accumulating in the trenches was recognized in 1963 when the center's operating permit was issued. With the continual rise of water in three of the northern trenches, the permit was revised in 1968 to include a construction design for future trenches to minimize infiltration (Kelleher, 1979). The low permeability of the till at the site, combined with poor stability of the trench covers, continued to result in an accumulation of water in the trenches. State monitoring of the adjacent streams indicated a slight increase in the levels of ^{3}H in the water during the early 1970s, although a state-ordered study in 1973–74 (Scholle, 1983) showed that no significant migration of radioactivity from the trenches had occurred. Therefore, it is questionable whether this ^{3}H even came from the LLRW area. Although the problems associated with the water migration did not pose any health risks, when water seeped through the cover of two of the trenches in the northern area in 1975, political pressure led the operator to close the site voluntarily. No outside commercial waste was accepted at the site after that point (Dayal et al., 1986). Remedial action programs, which included pumping the leachate out of the trenches and recompacting the trench covers, were in effect from 1975 to 1981. After completing a New York State Department of Environmental Conservation (NYSDEC)-authorized rehabilitation program for pumping the trenches in 1980–81, significant remedial actions at the LLRW burial site were halted. However, environmental monitoring continues. This monitoring indicates that the water level in most of the trenches has risen since 1981 (Anderson, 1988). Sampling of gases escaping from the trenches provides evidence of the release of radioactive methane, carbon dioxide, and water (Fakundiny, 1985). Through this pathway, approximately 90% of the ^{14}C and as much as

25% of the ^{3}H buried in the trenches is expected to be released to the atmosphere within 10 to 15 years of its burial (Matuszek, 1988).

In February of 1982, the DOE assumed responsibility for the 3345-acre site and set up the West Valley Demonstration Project (WVDP), which is responsible for the solidification of DOE high-level liquid wastes stored at the Center as well as the decommissioning of the facilities. In the spring of 1983, the New York State Energy Research and Development Authority (NYSERDA) was officially given responsibility for the closed commercial burial site (Anderson, 1988). Since then the state has spent approximately $300,000 on remediation. For example, in 1986–87, water accumulation in northern trench No. 14 led NYSERDA to design and build a subterranean wall between the trench and a sandy layer that intersected the trench to stop the influx of water. The sand layer was then excavated and replaced with impermeable silty clay. This remediation appears to have corrected the influx problem. A number of regular maintenance services for the commercial site are provided by the DOE and West Valley Nuclear Services Company, which operates the nearby WVDP (Anderson, 1988).

Although New York State has banned the future use of the West Valley LLRW facility for commercial waste, both the United States and New York State Geological Surveys have reported that multiple sites within the 3345-acre complex are geohydrologically sound for use as LLRW disposal sites (Eisenbud, 1987; Fakundiny, 1985; NYSDEC 1987a, b, d, e; Prudic and Randall, 1979; Scholle, 1983). In fact, until 1986, the West Valley Nuclear Services Company used the NRC-licensed area for burial of LLRW generated from the HLW demonstration project. The past failures appear to be due more to the poor management practices of NFS and the generators (Matuszek, 1988) than any faults inherent in the site. Even though water analyses indicate that during heavy rainfall some contamination of the land surface occurred, this contamination did not pose any health hazards. Because this contamination is directly related to surface water drainage, reducing topsoil erosion or creating traps for fine sediments should have a great impact on reducing the possibility of radioisotope migration off-site via streams (Matuszek et al., 1979). Dose estimates from drinking water downstream from this site have been on the order of 0.02 mrem/yr (0.00002 mSv/yr) (Matuszek, 1988), less than one ten-thousandth of the average background dose received in the United States. A recent health impact study of the population in the region surrounding the West Valley site from 1973 to 1983 found no evidence of increased cancer incidence (Byers and Vena, 1986). Although the study was based on a small sample size, other epidemiological studies conducted by the New York State Health Department have failed to show evidence of any ill effects from the West Valley site.

Sheffield, Illinois

Approximately 90,500 m^3 of LLRW was disposed of during the 11 years of operation of this site before its closure in 1978. Although opened by California Nuclear, Inc., in 1968, US Ecology acquired the license and rights to the site and now owns approximately 300 acres around the 20-acre burial site, which is deeded to the state (NYSDEC, 1987c). Within the 320-acre area, US Ecology also operated two hazardous waste disposal sites. US Ecology owns the extended acreage because they were forced to buy all the land that sat over that part of the regional aquifer that was shown to be contaminated (primarily by ^3H). In 1974, US Ecology submitted a timely application for a permit for expansion. However, the NRC failed to act upon the application before site capacity was reached. With no capacity to accept additional waste and without permission to expand, in 1979 US Ecology tried to terminate, unilaterally, their license with the NRC and their lease with Illinois. The NRC and the Illinois State Department of Labor initiated litigation, claiming that US Ecology must obtain a license amendment to allow for closure. The NRC's Atomic Safety Licensing Board ruled in 1987 that US Ecology could not terminate its license and, moreover, that it was required to close the site according to the provisions of 10 CFR Part 61. The licensing board upheld this decision in an appeal brought by US Ecology. When Illinois became an agreement state in 1987, it inherited the ongoing legal battle with US Ecology, so that US Ecology is currently in litigation with the state over its attempt to terminate both the license and the lease. The state is trying to settle out of court to allow for stabilization of this site, which has not received any waste since 1979 and yet still does not have permanent caps over the trenches. Until the litigation is solved, US Ecology must abide by the regulations as though it had a license and remains responsible for the maintenance and environmental monitoring of the facility (personal communication with David Ed, Illinois Department of Nuclear Safety).

The major problems at Sheffield are associated with erosion and trench subsidence, as well as the detection of radioactivity in on-site monitoring wells (Dayal et al., 1986). However, the trenches are in well-drained systems so that the infiltrated water has a relatively short residence time and is not continually present in the trenches, greatly reducing the potential for leaching and microbial degradation of the waste (Dayal et al., 1986). For drums that were breached upon emplacement, all the ^{3}H is assumed to have escaped during the first year after burial. For those that are not breached, approximately 4.5% of the ^{3}H escapes per year (MacKenzie et al., 1985).

Migration of water from the burial trenches is the primary problem that led to the closure of these three sites. At the West Valley and Maxey Flats sites, high annual precipitation, permeable trench covers, and impervious substrata caused collection of water in the trenches and a subsequent overflow of contaminated, laterally migrating water. At Maxey Flats, ^{3}H in the groundwater suggests that some subsurface migration occurred as well, while contamination from evaporation of leachate from the trenches contributed to off-site doses. These sites have since undergone extensive remedial actions designed to restabilize the trenches and improve drainage; trenches have also been capped with heavyweight plastic sheeting (NLLRWMP, 1984). Despite these efforts, site drainage continues to pose problems, requiring constant monitoring. Periodic pumping of trenches to remove the accumulations of leachate continues at Maxey Flats, but not at West Valley or Sheffield. The water infiltration problem is exacerbated by the poor waste management practices implemented during the early years of operation. For instance, the steel drums used for burial containers rust within a year of burial in a wet trench (Kempf et al., 1987). The permeation of the containers, especially those containing nonstabilized waste, leads to problems with compaction and subsidence, microbial degradation, and release of gases. However, only at the Maxey Flats site was the potential for health risk the reason for closure of a site, and even that claim was discredited by the EPA. The decision to close West Valley appears to have been a politically motivated one, not based on any real threat to the public's health (Matuszek, 1988; Spath and Hornibrook, 1986), while the closure of the Sheffield site resulted from inaction on the part of the NRC on US Ecology's petition for expansion.

Operating Sites

With the closure of three sites between 1975 and 1979, the burden of disposal of the country's LLRW has fallen on the three remaining sites. Although the volume disposed of annually has decreased (Figure 1-13), approximately 53,000 m^3 was received at the sites in 1987: 51.8% of the volume and 78.3% of the activity at Barnwell; 30.2% of the volume and 17.6% of the activity at Richland; and 18% of the volume and 4% of the activity at Beatty. In 1988, the three sites accepted a total volume of 40,472 m³ (NLLRWMP, 1988; Helminski, 1989a).

Barnwell, South Carolina

This site is owned by the state of South Carolina and operated by Chem-Nuclear Systems, Inc. (Figure 2-2). The facility, located adjacent to the DOE's Savannah River nuclear power research and waste disposal complex, occupies approximately 300 acres, using 47 acres for disposal, with an additional 160 acres allotted to future disposal (Ebenhack, 1983). Since its opening in 1971, despite the temperate climate and high annual precipitation rate, no significant problems have arisen, and this site continues to accept the largest percentage of LLRW generated in the country. Success at keeping trenches dry appears to be due to the fact that the soil underlying the Barnwell trenches is at least one-thousand-fold more porous than at Maxey Flats or West Valley. Infiltrating water simply drains out the bottom of the trench. Approximately 75% of the volume of waste accepted by Chem-Nuclear is derived from the nuclear power industry, with

FIGURE 2-2. Aerial view of the Barnwell, South Carolina, LLRW disposal facility. This site began commercial operation in 1971 and is one of the three remaining viable sites. *(Courtesy of Chem-Nuclear Systems, Inc.)*

the remaining volume coming from non-fuel-cycle generators (Dayal et al., 1986). Through 1987, the site, with a projected total capacity of 975,000 m^3, had accepted 559,572 m^3 of waste (NLLRWMP, 1987, 1988). The LLRW is buried in either narrow slit trenches or in standard SLB trenches (Dayal et al., 1986). Although waste packages are now generally stacked in the trenches, in the past drums and other small boxes were typically dumped in to fill the spaces between and alongside the other wastes (Clancy et al., 1981).

Despite its record of accepting large volumes, South Carolina has made it clear on several occasions, for example in 1979, that it does not wish to be the repository for the entire country or for all types of LLRW. After the Three Mile Island incident in 1979, the governor of South Carolina refused to allow the Barnwell site to receive any of the waste that resulted from the accident. TRU waste in concentrations greater than 10 μCi/g (370 kBq/g) has never been accepted (Dayal et al., 1986), nor have mixed wastes even if absorbed or solidified. The ash from incineration of such wastes is accepted only if solidified (Bowerman et al., 1986). In November 1979, Chem-Nuclear implemented a plan that cut in half the amount of waste accepted at the site by 1981. Although the site will not reach capacity at the current rate of waste acceptance until 2009, Chem-Nuclear plans to continue operating this site only until the end of 1992.

Because of numerous violations of federal LLRW transportation regulations by shippers and waste generators, the state of South Carolina instituted inspection procedures with strict penalties for violations, including the possible suspension of the violator's disposal license (NYSDEC, 1987c). Chem-Nuclear performs random inspections of waste to verify compliance with the free-liquid requirement of 10 CFR Part 61 and uses a customer compliance program to screen out mixed wastes (Bowerman et al., 1986).

Richland, Washington

The LLRW disposal site at Richland has been in operation since it was opened in 1965 by California Nuclear, Inc. (Figure 2-3). In 1968, Nuclear Engineering Company, now called US Ecology, acquired the site (NYSDEC, 1987c). For a short time in the fall of 1979, the governor ordered the site closed, blaming faulty packaging and federal transportation violations, sending the same message as the governor of South Carolina. By the end of 1987, Richland, with its total capacity of 1,494,000 m^3 had accepted 294,234 m^3 of waste (NLLRWMP, 1987, 1988). The site is operated by US Ecology, Inc., within a 1000-acre tract of land on the Hanford Nuclear Reservation, which the DOE leases to the State of Washington (Washington State Department of Social and Health Services, 1986). The state, in turn, subleases 100 acres of this land to US Ecology. The annual volume of waste received at this site steadily increased during the years 1970 to 1985. However, that trend has reversed since 1986. The volume disposed of in 1987 was only 83% of that disposed of in 1986, and that of 1986 only 47% of what was disposed of in 1985 (Ko, 1988). Of all the waste accepted, 98% is Class A waste. The concentration of the Class A waste has increased 98% since the first supercompactor came on-line, from an average of 148.3 mCi/m³ (5.59 GBq) in 1984 to 293.1 mCi/m³ (10.8 GBq) in 1987 (Ko, 1988).

Most of the waste is buried in conventional trenches. However, the site also contains 4 caissons (30-ft-deep vertical wells, 2 ft in diameter) of high-activity wastes, a hazardous chemicals trench, underground solar evaporation tanks (which flooded in 1984), and a buried reactor head in the special projects area. The maximum size of any one of the 18 trenches is 1000 ft long, 150 ft wide, and 45 ft deep (Carlin and Hana, 1988).

a

FIGURE 2-3. (a) Aerial view of the Richland, Washington, LLRW disposal facility. The site opened in 1965 near the DOE's Hanford Reservation; (b) emplacing LLRW in an SLB trench at the Richland site.

Unlike the management practices at the other two operating sites, US Ecology does not use liners in the trenches at the Richland site, nor do they perform any monitoring of the air or trench water. With no on-site monitoring wells, no data on water migration or contamination are available. The trenches do not have leachate covers, and no means of detecting water accumulation or gas emission exists. However, like the Beatty, Nevada, site, Richland is located in an arid region of the country. A number of other aspects of past management practices have contributed to uncertainties about the proper maintenance program and the type of postclosure care that the facility will require. The nature of the buried waste has not been adequately documented and cannot be verified. Neither a site characterization nor

b

measurement of background radiation levels was performed prior to commencement of the disposal operations; environmental monitoring programs were not started until 1980; and before 1979, the state performed only cursory, prescheduled annual inspections of this site (Carlin and Hana, 1988).

Following EPA waste disposal regulations, the state of Washington banned the disposal of scintillation fluids in 1985, and the disposal of activated lead in 1987, although activated lead used as shielding is still accepted. US Ecology does accept some nonhazardous, that is, not Resource Conservation and Recovery Act (RCRA)-listed, organic liquids if solidified or absorbed; waste containing organic liquids with 0.05 μCi/g (1.85 kBq/g) or less of ^{3}H or ^{14}C is not accepted (Bowerman et al., 1986). US Ecology does not routinely inspect for mixed wastes, but will do so if requested by a state regulatory agency provided that the radiation hazard is not greater than 10 R/hr (2.58 $\times$ 10^3 C/kg) on contact (Bowerman et al., 1986).

The Richland site will also serve as the disposal facility for the Northwest compact. Out-of-compact waste will continue to be accepted through 1992. After that time, only waste from the compact states, an anticipated 2832 m^3/yr, will be accepted at Richland (Ko, 1988).

Beatty, Nevada

US Ecology, Inc., has operated this site since 1962. The state temporarily suspended the operating license in March, 1979, when it was discovered that workers had been removing contaminated tools from the site (Wenslawski and North, 1979). In the fall of 1979, the governor of Nevada, like those of Washington and South Carolina, ordered the site closed temporarily due to problems with packaging. The state tried to close the site permanently in 1980 and implemented regulations requiring third-party inspections of all shipments to the site. Beatty accepts the smallest percentage of LLRW of all three operating sites, only 5% of the total volume in 1986. Through the end of 1987, the site had accepted 110,202 m^3 of waste, and expects to be able to accept another 27,798 m^3 (NLLRWMP, 1987, 1988). No significant problems have developed from the disposal of LLRW at this site. The waste facility occupies approximately 47 acres within an 80-acre area leased from the state of Nevada. The site also has a hazardous chemical waste disposal facility, separated from the LLRW disposal site by a 10-acre buffer zone. Both sites are surrounded by a 400-acre state-leased buffer zone (NYSDEC, 1987c). Beatty, in the desert of Nevada, is ideally located with respect to geology and climate. Until the Rocky Mountain Compact develops a new site in Colorado, projected to be completed in 1992, Beatty will serve as the disposal facility for the compact (NYSDEC, 1987c).

Sanitary Landfills

All regulated LLRW generated in this country is being disposed of at either one of the three commercial sites or a DOE site. In Texas, however, some of its LLRW has been deregulated and may now be buried in sanitary landfills. In 1987, the Texas Low-Level Radioactive Waste Disposal Authority successfully petitioned the Texas Department of Health to establish levels "below regulatory concern" (BRC) based on the maximum permissible concentrations in air and water for radioisotopes with half-lives of less than 300 days. At a Type I sanitary landfill, municipal waste is compacted and covered with at least 6 in. of soil daily, the trenches have low-permeability liners, and access to the site is restricted (Rogers and Murphy, 1987). Texas generators may now request license amendments from the state Bureau of Radiation Control, which allow for disposal of limited concentrations of short-lived radioisotopes in Type I sanitary landfills. This ruling allows for disposal of the BRC materials as municipal waste, possibly transported and handled by nonradiation workers, and buried or incinerated at the landfill without regard to its radioactive content. This ruling met with very limited opposition from the Texas public, apparently due to the strong sense of public responsibility for management of waste produced by the state, a large amount of community relations, and state pride.

Treatment Centers

The business of LLRW management is not limited to the burial sites. A number of treatment sites exist that serve as intermediaries between the generators and the burial site operators. In addition to the brokers who transport LLRW, there are several commercial compaction facilities. Scientific Ecology Group (SEG) in Tennessee, a subsidiary of Westinghouse, has operated a supercompactor facility since 1987. During its first year, SEG compacted 14,925 m^3. At present, SEG accepts approximately 90% of all commercial nuclear power reactor and DAW LLRW in the country (in 1988, SEG accepted over

28,321 m³; in 1989, SEG anticipated accepting 33,985 m³) for treatment, dealing with over 80% of the brokers and approximately 70% of all industrial generators in the United States (Figure 2-4).

RAMP is a new supercompactor center in Laurel, Maryland. Chem-Nuclear, Inc., operates a supercompactor in Shannahan, Illinois. Allied Technology runs a supercompactor facility in Richland, Washington. US Ecology and Westinghouse have mobile supercompactor units that are used primarily at reactor sites. When not in use elsewhere, US Ecology operates its mobile supercompactor at a permanent site in Memphis. After compaction, the waste is shipped to one of the three disposal sites.

Both SEG and RAMP applied for permits to operate an incinerator for LLRW. SEG received all necessary federal, state, and local permits in late 1988. These permits include an EPA radiological air quality permit, a state air pollution control permit, a state radiological license, and permission from Oak Ridge city council to install the incinerator equipment (Helminski, 1988d). The incinerator began operation in November 1989, with metal-melting capabilities to be added in 1990. SEG will accept all Class A waste, compacted or not, for incineration,

FIGURE 2-4. A commercial treatment center for LLRW. Here, SEG supercompacts LLRW from generators located throughout the country. Compacted materials are subsequently repackaged for burial. (*Courtesy of Scientific Ecology Group.*)

but they would prefer to receive only materials with low PVC and carbon content, and no high-activity ^{3}H wastes. They are planning to vitrify the ash. The combination of compaction, incineration, and vitrification will provide a 40- to 450-fold volume reduction. SEG is planning to expand its facilities to accept mixed wastes beginning in 1990 or 1991 (personal communication, Gary Lester, SEG).

Although closed now, Todd Shipyard in Texas, an abandoned facility, was used for storage and decay of commercial LLRW. Because of complaints about the odor and appearance of the site, the state requested that the operators reduce the amount of waste being held and improve general management. The operator refused and the state closed the site in 1979 (Colglazier and English, 1988).

The Quadrex Recycling Center, Florida, receives the bulk of the nation's deregulated scintillation vials containing *de minimis* levels of ^{3}H and ^{14}C, as well as other radioisotopes at a concentration of 0.002 μCi/g (0.074 kBq) or less (Figure 2-5). At the center, the vials are crushed and the solvents are transferred to the Oldover Cement Company, which uses them as fuel in a rotary kiln. Although the glass and plastic waste was originally sent to a municipal landfill, complaints of toluene odor from the site led Quadrex to send the waste glass and plastic to an industrial landfill. When Quadrex began processing liquid scintillation vials in 1983, the process was exempt from certain RCRA requirements. In 1986, the EPA began requiring permits for the storage of flammable hazardous waste under the RCRA "Fuel Marketeer" rule. EPA granted Quadrex a permit in July 1988 that is valid for 3 years, after which time Quadrex will need to renew its permit (Helminski, 1988a). There are now several other companies that accept scintillation vials, for example, Thermalkem in South Carolina and TWI in Missouri.

Quadrex also operates a decontamination center in Oak Ridge, Tennessee, where reactor materials are treated by sandblasting and solvent washing. In May 1989, the Tennessee Department of Health and Environment announced that waste processors in Tennessee, for example, SEG and Quadrex, would be prevented from processing, storing, or treating any LLRW from states that are not in compliance with the Amendments Act (Helminski, 1989b).

The NRC, in setting criteria for LLRW disposal site development, has approved the use of SLB, stating that this method can adequately meet their safety standards. The history of the sites illustrates this point. Although some have experienced water problems and small amounts of radioactivity have escaped into surface or groundwaters, the only measurable human exposure occurred at the Maxey Flats site. This exposure resulted in an annual dose to a few people of less

FIGURE 2-5. The Quadrex HPS treatment facility, located in Gainesville, Florida, processes deregulated liquid scintillation counting (LSC) wastes. (*a*) Drums of LSC vials are prepared for processing; (*b;* opposite) drum contents are emptied into a hopper and then crushed. Liquids are drained and collected while solids are washed; (*c*, p. 66) bulk LSC fluids are pumped into a tanker truck and transported to a rotary cement kiln for incineration; (*d*, p. 66) drums of crushed glass and plastics awaiting disposal as nonradioactive waste. (*Courtesy of Quadrex HPS, Incorporated.*)

than 1 mrem (0.01 mSv) (Eisenbud, 1988), well within the limits of 10 CFR Part 61 despite the fact that the site was built long before Part 61 was put into effect.

Numerous studies of the commercial burial sites are available and should be consulted by the developers of new sites for guidance in establishing selection and development criteria, particularly by those groups considering SLB. The most important lesson learned is that, despite proof that SLB is safe, problems resulting from the poor waste management practices of the past, such as trench subsidence, cast doubt on the adequacy of this method in the minds of an already distrusting public. Proper waste management could eliminate or reduce these problems, calm unjustified fears, and lower operating costs at

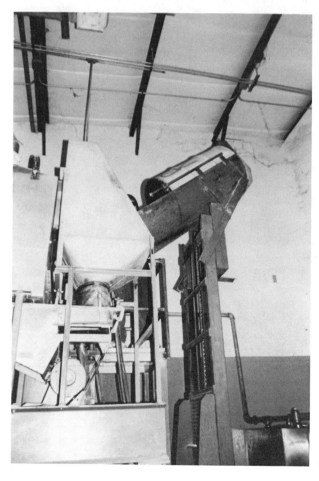

b

**FIGURE
2-5c**

**FIGURE
2-5d**

the sites. The wastes should be segregated by their physical and radioactive characteristics, properly packaged and placed, not simply dumped, into the trenches to reduce subsidence and increase stability. The trenches should be located in well-drained systems in order to reduce the residence time of water in the trenches. Considering that the integrity of a steel drum in a wet trench is often breached in less than one year, the use of noncorrodible containers would greatly reduce the chance of drum failure and the resulting compaction. Another lesson learned from the closed sites is that the cost of postoperational monitoring is higher than anticipated, and all future sites should account for the possibility of decommissioning at any time during operation (Mallory, 1983). For example, the Washington State Department of Ecology estimates that closure of the Richland site will cost $14 million and perpetual care and maintenance will cost an additional $23 million. At present the state has only $16 million in the postclosure trust fund (Helminski, 1988c).

Disposal of LLRW in Other Countries

In most European countries, LLRW is now disposed of in SLB facilities. As was common in the United States, LLRW was buried at sea until 1982. In France, the Commissariat à l'Energie Atomique (CEA), through L'Agence Nationale pour la Gestion des Déchets Radioactifs (ANDRA), regulates the disposal of radioactive materials. The national disposal site, presently operating at Centre de la Manche near La Hague, opened in 1969 and is anticipated to operate through the 1990s. The French employ earth-mounded concrete bunkers for disposal. ANDRA is developing a new site at Centre de l'Aube that is to have a 30-year life.

The Swedish Nuclear Fuel Supply Company (SKB) began operating the SFR repository in 1988, using it for the disposal of both LLRW and intermediate-level radioactive waste. The repository is located in crystalline rock, a few hundred meters offshore of Forsmark, 50 m below the seabed. It is accessed by an inclined tunnel from the coast (Chapman and McKinley, 1987).

The Federal Republic of Germany (FRG) disposes LLRW in an abandoned iron mine at the Konrad site. All LLRW is collected for central treatment within each federal state. No distinction is made between utility and non-utility waste (Kuehn, 1989).

In Great Britain, the main LLRW disposal site at Drigg, in Cumbria, is operated by and for British Nuclear Fuels plc (BNFL), and uses standard SLB. Since 1959, this site has accepted power plant wastes and LLRW from the National Disposal Service (NDS), which handles non-power-plant wastes.

Great Britain is searching for a new site. In 1986, the Department of the Environment, which governs the handling of radioactive materials, selected a group of potential sites (Bradwell, Essex, Elstow, Bedfordshire, Fulbeck, Lincolnshire and Killingholme, Humberside) for a new near-surface facility and site evaluations began (Chapman and McKinley, 1987). But in May 1987, the Nuclear Industry Radioactive Waste Executive (NIREX) announced that, due to cost considerations, LLRW should be disposed with the intermediate-level waste (ILW) in a deep repository (Wilkinson, 1989).

By 1998, Switzerland should have operational a mined repository for all LLRW. The facility will be a horizontal tunnel connected to a cavern system for higher-activity wastes mined out of marl (at Oberbauenstock), anhydrite (at Bois de la Glaivaz), or crystalline rock (at Piz Pian Gran) (Chapman and McKinley, 1987).

Finland is developing a repository for LLRW in crystalline rocks that will be similar to the Swedish SFR facility. This repository should be operational in 1992.

The National Agency for Radioactive Waste and Fissile Materials (ONDRAF-NIRAS) in Belgium provides centralized treatment and conditioning for LLRW. Until 1982, all LLRW was dumped into the ocean. Since 1983, it has been stored in aboveground vaults at the Mol-Dessel site awaiting the development of an acceptable land disposal facility (Detilleux et al., 1989).

In Italy, until 1982, all generators treated and held their own LLRW. In that year, a public company, NUCLECO, began collection, treatment, and disposal services for LLRW generators.

The Atomic Energy Commission of Japan discontinued ocean dumping in 1982. The Japanese are now planning an SLB facility at Rokkasho Mura, Aomore Prefecture, which is scheduled to begin operation in 1991. Until this facility opens, all generators are storing their own LLRW (Kiyose and Tsunoda, 1989).

The Atomic Energy Council of Taiwan has authorized the incineration of power plant LLRW beginning in 1989 or 1990. Non-power-plant waste is segregated for storage. Previously, Taiwan ocean-dumped its LLRW. Currently all LLRW is stored awaiting solidification offshore on a small island. No decision on a permanent solution for disposal has yet been made (Soong and Liu, 1989).

Although in the past, Canada disposed of LLRW in SLB sites or mounded bunkers, the Atomic Energy Control Board now authorizes only interim storage of LLRW. No disposal is performed. The board is currently evaluating disposal methods.

Transportation of LLRW

All LLRW is shipped to the burial facilities by ground transportation, regulated by the U.S. DOT under 49 CFR Parts 100-179. In the rare occurrence of a transportation accident involving LLRW, the Federal Emergency Management Agency, along with eleven other agencies, has established an emergency response plan (*Federal Register*, 1984, 1985). The DOT regulations address third-party bodily injury and property damage and cleanup. They require minimum limits for insurance of $1,000,000 per occurrence and $5,000,000 per occurrence for route-controlled highway shipments. All transporters of LLRW must maintain full compliance with the Motor Carrier Act of 1980 (Carlin and Hana, 1988). In cases where states, such as New York, have tried to prohibit or restrict movement of these radioactive materials, the federal government has invoked the Commerce Clause to override the state regulations.

Approximately 1 in every 250,000 shipments of commodities in the United States involves radioactive materials of some sort. In 1984, only 24 of the 5956 hazardous materials incident reports filed with the DOT involved radioactive materials, and "reviews of available historical data have shown that there has never been a serious [transportation] incident involving the dispersal of radioactive material" (IAEA, 1986; Murray, 1989).

References

Anderson, D. B. 1988. Surveillance and maintenance of the West Valley state licensed low-level radioactive waste disposal area 1983–1987. *Waste Management '88* 1:643–650.

Bowerman, B. S., R. E. Davis, and B. Siskind. 1986. *Document Review Regarding Hazardous Chemical Characteristics of Low-Level Waste.* NUREG/CR-4433. Upton, N.Y.: Brookhaven National Laboratory.

Bowerman, B. S., C. R. Kempf, D. R. MacKenzie, B. Siskind, and P. L. Piciulo. 1985. *An Analysis of Low-Level Wastes: Review of Hazardous Waste Regulations and Identification of Radioactive Mixed Wastes.* NUREG/CR-4406, Springfield, Va.: NTIS.

Byers, T., and J. Vena. 1986. Cancer incidence in the region around West

Valley, New York, 1973–1983. Unpublished report for the Coalition on West Valley Nuclear Wastes, 15 December 1986.

Carlin, E. M., and S. L. A. Hana. 1988. Impact of liability and site closure and long-term care issues on future siting efforts. *Waste Management '88* 1:61–67.

Carter, M. W., and D. C. Stone. 1986. Quantities and sources of radioactive waste. *Proceedings of the 21st Annual Meeting of the National Council on Radiation Protection and Measurements.* Washington, D.C.: NCRP, pp. 5–30.

Chapman, N. A., and I. G. McKinley. 1987. *The Geological Disposal of Nuclear Waste.* Chichester, Great Britain: John Wiley & Sons.

Clancy, J. J., D. F. Gray, and O. I. Oztunali. 1981. *Data Base for Radioactive Waste Management: Review of Low-Level Radioactive Waste Disposal History.* NUREG/CR-1759. Washington, D.C.: NRC.

Colglazier, E. W., and M. R. English. 1988. Low-level radioactive waste: Can new disposal sites be found? In *Low-Level Radioactive Waste Regulation: Science, Politics, and Fear,* ed. M. E. Burns. Chelsea, Mich.: Lewis Publishers, pp. 215–238.

Dayal, R., R. F. Pietrzak, and J. H. Clinton. 1986. *Geochemical Studies of Commercial Low-Level Radioactive Waste Disposal Sites.* NUREG/CR-4644. Washington, D.C.: NRC.

Detilleux, E., F. Decamps, and R. Heremans. 1989. Radioactive waste management activities and related research in Belgium. *Waste Management '89* 1:49–53.

Ebenhack, D. G. 1983. Barnwell low-level waste disposal operations. In *The Treatment and Handling of Radioactive Wastes.* eds. A. G. Blasewitz, J. Davis, and M. R. Smith. Columbus, Ohio: Battelle Press, pp. 460–461.

Eisenbud, M. 1987. *Environmental Radioactivity.* 3rd ed. Orlando, Fla.: Academic Press.

Eisenbud, M. 1988. Low-level radioactive waste repositories: A risk assessment. Paper read at the Annual Conference of the North Carolina Academy of Science, 26 March 1988, Charlotte, N.C.

Fakundiny, R. H. 1985. Practical application of geological methods at the West Valley low-level radioactive waste burial ground, Western New York. *Northeastern Environmental Science* 4:116–148.

Federal Register. 1984. 49:35896.

Federal Register. 1985. 50:46542.

Godbee, H. W., and A. H. Kibbey. 1983. State-of-the-art review on the management of low-level radioactive transuranic wastes. In *The Treatment and Handling of Radioactive Waste,* eds. A. G. Blasewitz, J. M. Davis, and M. R. Smith. Columbus, Ohio: Battelle Press, pp. 347–351.

Health and Environment Network. 1988. Leaking LLRW means costly cleanup. *Health & Environment Digest* 2:5.

Helminski, E. L. 1988a. Florida issues hazardous waste permit to Quadrex. *Radioactive Exchange* 7(13):3.

Helminski, E. L. 1988b. US Ecology sues Kentucky over Maxey Flats clean-up costs. *Radioactive Exchange* 7(13):3–4.

Helminski, E. L. 1988c. Wrap up (LLRW): In the northwest. *Radioactive Exchange* 7(18):5.

Helminski, E. L. 1988d. SEG incinerator receives final state, federal, local approval. *Radioactive Exchange* 7(21–22):2–3.

Helminski, E. L. 1989a. LLRW volume disposal update. *Radioactive Exchange* 8(1):8–9.

Helminski, E. L. 1989b. Tennessee bans processing of LLRW from non-compliance states. *Radioactive Exchange* 8(10):1–2.

International Atomic Energy Agency (IAEA). 1986. *Assessment of the Radiological Impact of the Transport of Radioactive Materials.* Technical document #398. Vienna, Austria: IAEA.

Kelleher, W. J. 1979. Water problems at the West Valley burial site. In *Management of Low-Level Radioactive Waste,* eds. M. W. Carter, A. A. Moghissi, and B. Kahn. New York: Pergamon, pp. 843–851.

Kempf, C. R., C. J. Sven, S. Mughabghab, and T. M. Sullivan. 1987. *Low-Level Waste Source Term Evaluation: Quarterly Progress Report, January–March 1987.* WM-3276-2. Upton, N.Y.: Brookhaven National Laboratory.

Kiyose, R., and N. Tsunoda. 1989. Progress in radioactive waste management in Japan. *Waste Management '89* 1:25–29.

Ko, S. 1988. Analysis of the reduction in waste volumes received for disposal at the low-level radioactive waste site in the state of Washington. *Waste Management '88* 1:629–635.

Kuehn, K. 1989. U.S.-German cooperation in radioactive waste disposal: Are the results of ten years' experience a solid basis for the future? *Waste Management '89* 1:31–34.

Mallory, C. W. 1983. Alternative for future land disposal of radioactive waste. In *The Treatment and Handling of Radioactive Waste,* eds. A. G. Blasewitz, J. M. Davis, and M. R. Smith, Columbus, Ohio: Battelle Press, pp. 462–465.

Matuszek, J. M. 1988. Safer than sleeping with your spouse—the West Valley experience. In *Low-Level Radioactive Waste Regulation: Science, Politics, and Fear,* ed. M. E. Burns. Chelsea, Mich.: Lewis Publishers, pp. 260–278.

Matuszek, J. M., L. Husain, A. Lu, J. F. Davis, and R. H. Fakundiny. 1979. Application of radionuclide pathways studies to management of shallow, low-level radioactive waste burial facilities. In *Management of Radioactive Waste,* eds. M. W. Carter, A. A. Moghissi, and B. Kahn. New York: Pergamon, pp. 901–914.

Miller, W. O., and R. D. Bennett. 1985. *Alternative Methods for Disposal of Low-Level Radioactive Wastes: Task 2c: Technical Requirements for Earth Mounded Concrete Bunker Disposal of Low-Level Radioactive Waste.* NUREG/CR-3774, Vol. 4, Washington, D.C.: NRC.

Montgomery, D. M., H. E. Kolde, and R. L. Blanchard. 1977. Radiological measurements at the Maxey Flats radioactive waste burial site—1974 to 1975. Rep. EPA-520/5-76/020. Office of Radiation Progams, USEPA: Cincinnati, Ohio.

Murray, R. L. 1989. *Understanding Radioactive Waste.* 3d ed. Columbus, Ohio: Battelle Press.

National Council on Radiation Protection and Measurements (NCRP). 1987. *Public Radiation Exposure from Nuclear Power Generation in the United States.* NCRP No. 92, Bethesda, Md.: NCRP.

National Low-Level Radioactive Waste Management Program (NLLRWMP). 1984. *Experience and Related Research and Development in Applying Corrective Measures at Major Low-Level Radioactive Waste Disposal Sites.* DOE/LLW-28T. Springfield, Va.: NTIS.

National Low-Level Radioactive Waste Management Program (NLLRWMP). 1987. *The 1986 State-by-State Assessment of Low-Level Radioactive Wastes Received at Commercial Disposal Sites.* DOE/LLW-66T. Springfield, Va.: NTIS.

National Low-Level Radioactive Waste Management Program (NLLRWMP). 1988. *The 1987 State-by-State Assessment of Low-Level Radioactive Waste.* DOE/LLW-69T. Springfield, Va.: NTIS.

New York State Department of Environmental Conservation (NYSDEC). 1987a. *Final Environmental Impact Statement for Promulgation of 6 NYCRR Part 382: Regulation for Low-Level Radioactive Waste Disposal Facilities, Executive Summary.* New York: NYSDEC.

New York State Department of Environmental Conservation (NYSDEC). 1987b. *Final Environmental Impact Statement for Promulgation of 6 NYCRR Part 382: Regulation for Low-Level Radioactive Waste Disposal Facilities, Vols. I and II.* New York: NYSDEC.

New York State Department of Environmental Conservation (NYSDEC). 1987c. *Recommendations for State Assistance to Localities Affected by the Siting of a Low-Level Radioactive Waste Management Facility.* New York: NYSDEC.

New York State Department of Environmental Conservation (NYSDEC). 1987d. *Siting and Disposal Techniques, Draft 6 NYCRR Part 382: Regulations for Low-Level Radioactive Waste Disposal Facilities.* New York: NYSDEC.

New York State Department of Environmental Conservation (NYSDEC). 1987e. *6 NYCRR Part 382: Regulations for Low-Level Radioactive Waste Disposal Facilities.* New York: NYSDEC.

New York State Department of Environmental Conservation (NYSDEC). 1987f. *State Assistance to Local Governments.* Boston, Mass.: Energy Systems Research Group.

Prudic, D. E., and A. D. Randall, 1979. Groundwater hydrology and subsurface migration of radioisotopes at a low-level solid radioactive waste disposal site, West Valley, New York. In *Management of Low-Level*

Radioactive Waste, eds. M. W. Carter, A. A. Moghissi, and B. Kahn. New York: Pergamon, pp. 853–882.

Rogers, V. C., and E. S. Murphy. 1987. *Disposal of Short-Lived Radionuclide Wastes in a Sanitary Landfill.* Austin, Texas: Texas Low-Level Radioactive Waste Disposal Authority.

Scholle, S. R. 1983. An evaluation of Sierra Club concerns about subsurface migration of radioactive wastes at West Valley, New York. *Northeastern Environmental Science* 2:8–12.

Soong, K. L., and S. J. Liu. 1989. Current status of spent fuel disposal program in Taiwan, Republic of China. *Waste Management '89* 1:77–82.

Spath, J., and C. Hornibrook. 1986. *Planning Report on Management of Low-Level Radioactive Waste.* New York: NYSERDA.

Washington State Department of Social and Health Services. 1986. *Fact Book: Disposal of Low-Level Radioactive Waste in Washington State.* Olympia, Wash.: Washington State.

Wenslawski, F. A., and H. S. North, Jr. 1979. *Low-Level Radioactive Waste Management, Proceedings of the Health Physics Society Midyear Topical Symposium, 11–15 February 1978, Williamsburg, Virginia.* Published by the EPA as: EPA-520-3-79-002.

Wilkinson, W. L. 1989. The nuclear scene in the UK. *Waste Management '89* 1:35–39.

3 VOLUME REDUCTION AND WASTE PROCESSING

Volume reduction and waste processing are essential steps in any radioactive waste disposal program. These pretreatment methods bring immediate financial savings to the generator by reducing the volume of waste that must be shipped off-site. Furthermore, they reduce the surcharges levied against the volume of waste shipped. Reducing the amount of waste that must be buried also eliminates far-reaching "cradle to-grave" liabilities and preserves space for the disposal of materials that do require long-term isolation. For the site operator, smaller volumes of well-characterized material are easier to manage. Better management instills greater public confidence in the disposal technology, and reducing the volume of waste reduces the magnitude of the entire problem.

Financial and legal incentives for volume reduction have led nuclear power plants to develop more extensive processing and treatment procedures for low-level radioactive waste (LLRW) than other generators. The Amendments Act requires that the operators of all commercial nuclear power plants initiate volume-reduction measures. The first phase of these measures, completed between 1986 and 1989, was required to achieve a 17.5% reduction in volume for those plants in sited states or compacts and 30% for those in unsited states or compacts. Additional, more significant measures must be taken between 1990 and 1992—25 and 45% volume reduction for plants in sited and unsited compacts, respectively. As an example, Consolidated Edison's Indian Point Nuclear Station reduced its waste volume over 80% from 1600 m^3 in 1981 to less than 205 m^3 in 1987, and the total activity shipped for burial from over 2000 Ci/yr (74 TBq/yr) to several hundred

TABLE 3-1. Common Applications of Volume-Reduction
and Waste-Treatment Methods

Gases	Liquids	Sludges	Solids
Deregulation	Deregulation	Deregulation	Deregulation
Decay	Decay	Decay	Decay
Absorption	Sewer disposal	Sedimentation	Shredding
Ion exchange	Ion exchange	Centrifugation	Compaction
Scrubbing	Filtration	Filtration	Supercompaction
Filtration	Demineralization	Drying (forced-air)	Incineration
Liquification	Reverse osmosis	Linear dewatering	Pyrolysis
	Evaporation	Solidification	
	Crystallization	Incineration	
	Degradation	Pyrolysis	
	Reclamation	Compaction	
	Incineration	Supercompaction	

curies (Spall et al., 1988). This was accomplished by a combination
of waste-minimization and volume-reduction measures.

The first phase of the mandated volume-reduction measures has
been very effective. Similar legislation should be applied to other types
of generators of LLRW so that further reductions can be realized.
Institutions and some industries, especially those with research and
development laboratories, can also reduce their volumes of waste by
80% or more through a combination of segregation, decay of short-
lived isotopes, compaction, regulated sewer disposal, exclusion of
scintillation fluids, and incineration (NLLRWMP, 1987; Party and
Gershey, 1989). However, it is also important to consider that radiation
doses to workers at the disposal site may rise as a result of the
additional handling and environmental releases needed to carry out
these processes.

In order to identify the specific places where volume reduction is
possible and to select the appropriate methods, the physical and chemical composition of the waste stream must be well-characterized.
Table 3-1 gives an overview of volume-reduction methods while
Figures 3-1 and 3-2 show specific examples for nuclear power plants
and institutional generators of LLRW. The following is a brief summary of the various kinds of volume-reduction techniques available.
Although the steps outlined below have been instituted principally at
power plants, future emphasis on the reduction of all LLRW, regardless of source, should eventually encourage other generators to
follow suit.

Waste Minimization

Most important, generators should review the processes that generate their LLRW and efforts should be made to avoid or minimize waste production. Generators should use nonradioactive methods or short-lived radioisotopes wherever possible, maintain a clean work environment, avoid the unnecessary introduction of clean materials to contaminated areas, minimize the use of disposable protective equipment, operate an efficient maintenance program, and use easy-to-decontaminate materials and coatings (e.g., acrylic resin floor coverings) (IAEA, 1986b). Upgrading equipment and technology to improve operating efficiency often produces concomitant reductions in waste volumes.

Volume Reduction

While waste-minimization steps decrease the amount of LLRW, a number of different methods can be employed to reduce the volume that is still produced.

Segregation

Separation of wastes according to form (gaseous, liquid, solid), isotope, level of radioactivity, and chemical composition is the first step toward volume reduction. Establishment of and adherence to a uniform labeling system (Figure 3-3) that identifies the different kinds of wastes should help eliminate the troublesome and costly "unknown" category.

Decay

Waste containing short-lived radioisotopes should be stored for decay and then disposed of as ordinary trash (Figure 3-4). After 10 half-lives, only 0.1% of the original activity will remain. For a large fraction of LLRW, this 0.1% residue will be at or below background levels of radiation and can be disposed of without regard to its radioactivity. For wastes of high activity, this may not be practical since even after 10 half-lives, the remaining fraction may be too radioactive to dispose of without special handling. Depending upon the generator,

Boiling Water Reactor

TREATMENT

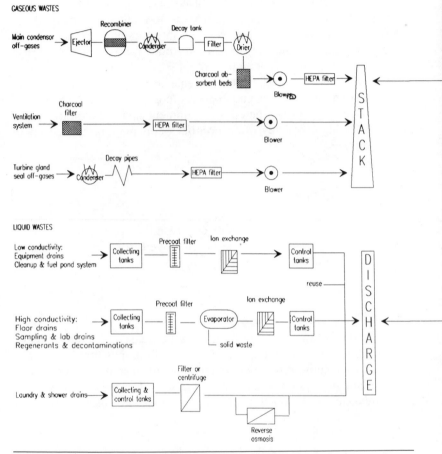

FIGURE 3-1. Specific treatment and volume-reduction methods for nuclear power plants. *(Adapted from IAEA, 1986.)*

Pressurized Water Reactor

TREATMENT

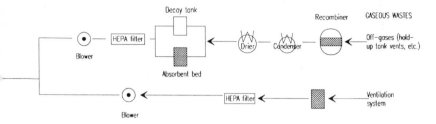

Decay tank

Recombiner   GASEOUS WASTES

HEPA filter

Blower

Drier   Condenser   Off-gases (hold-
up tank vents, etc.)

Absorbent bed

Blower   HEPA filter   Ventilation
system

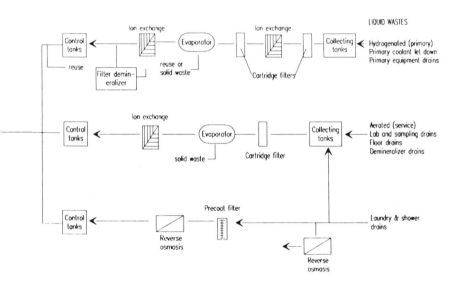

LIQUID WASTES

Control tanks   Ion exchange   Evaporator   Ion exchange   Collecting tanks   Hydrogenated (primary)
Primary coolant let down
Primary equipment drains

reuse   reuse or
solid waste   Cartridge filters

Filter demineralizer

Control tanks   Ion exchange   Evaporator   Collecting tanks   Aerated (service)
Lab and sampling drains
Floor drains
Demineralizer drains

solid waste   Cartridge filter

Control tanks   Precoat filter   Laundry & shower
drains

Reverse osmosis

Reverse osmosis

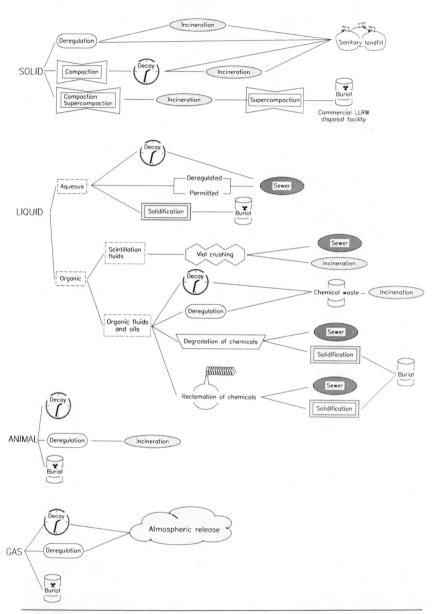

FIGURE 3-2. Specific volume reduction methods for laboratories. *(Party and Gershey, 1989.)*

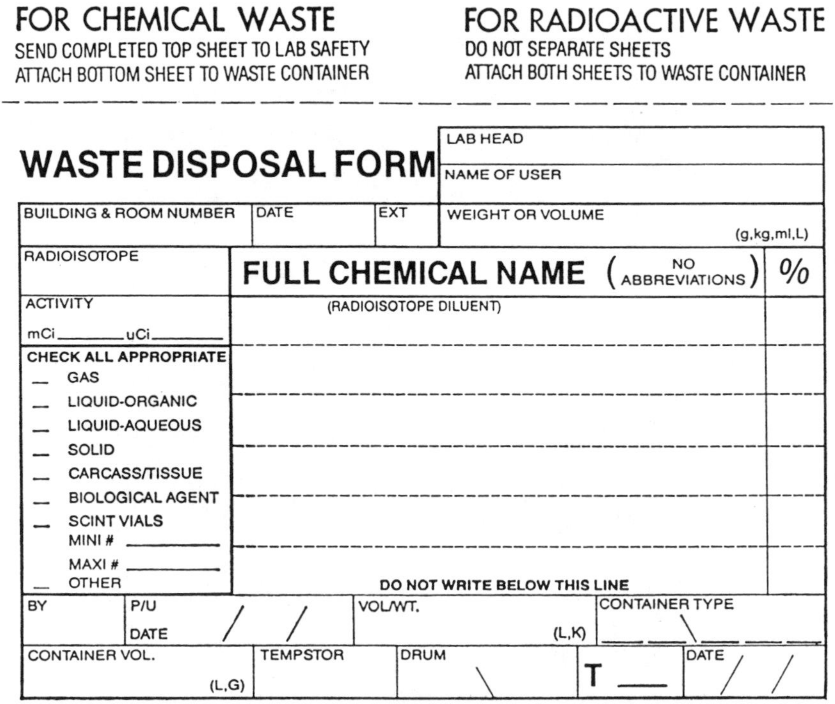

FOR CHEMICAL WASTE
SEND COMPLETED TOP SHEET TO LAB SAFETY
ATTACH BOTTOM SHEET TO WASTE CONTAINER

FOR RADIOACTIVE WASTE
DO NOT SEPARATE SHEETS
ATTACH BOTH SHEETS TO WASTE CONTAINER

WASTE DISPOSAL FORM	LAB HEAD
	NAME OF USER

BUILDING & ROOM NUMBER	DATE	EXT	WEIGHT OR VOLUME
			(g,kg,ml,L)

RADIOISOTOPE	**FULL CHEMICAL NAME** ($\begin{smallmatrix}NO\\ABBREVIATIONS\end{smallmatrix}$)	%
ACTIVITY	(RADIOISOTOPE DILUENT)	
mCi_____ uCi_____		
CHECK ALL APPROPRIATE		
__ GAS		
__ LIQUID-ORGANIC		
__ LIQUID-AQUEOUS		
__ SOLID		
__ CARCASS/TISSUE		
__ BIOLOGICAL AGENT		
__ SCINT VIALS		
MINI # _____		
MAXI # _____		
__ OTHER	DO NOT WRITE BELOW THIS LINE	

BY	P/U / /	VOL/WT.	CONTAINER TYPE
	DATE	(L,K)	
CONTAINER VOL.	TEMPSTOR	DRUM	DATE / /
(L,G)		T __	

FIGURE 3-3. Labeling system for the segregation and identification of LLRW.

a decay program can eliminate up to 70% of the waste shipped for burial. For generators using very short-lived radioisotopes, such as medical clinics, this method could altogether eliminate the need to ship LLRW. In general, it is practical to hold radionuclides with half-lives of up to 90 days on-site. If larger amounts of storage space were available, generators could hold wastes of longer-lived radioisotopes, perhaps those with half-lives up to 5 years. This would require greater attention to proper waste management, such as better packaging and monitoring.

Deregulation

A major portion of the Class A waste now shipped for burial could be eliminated if the regulatory agencies would establish levels for deregulation of materials contaminated with radioisotopes of half-lives greater than 90 days. By just the deregulation of ^{3}H and ^{14}C, insti-

FIGURE 3-4. An inexpensive, prefabricated structure used for on-site decay of low-activity, short-lived LLRW.

tutional generators could eliminate most of the waste that remains after separating out the decayable materials. [See Chapter 1 for a discussion of below-regulatory-concern (BRC) petitions.]

Deregulation by the Nuclear Regulatory Commission (NRC) of scintillation liquids containing insignificant amounts of ^{3}H and ^{14}C eliminated what had been the major component of the mixed waste produced in the United States. Because these liquids are no longer regulated by the NRC, they can be treated according to their composition and chemical hazard. Crushing the scintillation vials (Figure 3-5) to recover the liquid allows for bulk handling and disposal and provides additional volume reduction. Using nonflammable instead of

FIGURE 3-5. One possible volume-reduction method, a crusher (Balcan Engineering Ltd., Lincolnshire, England) for liquid scintillation vials. After crushing, the contents of the 55-gal (209-L) drum in the foreground will produce 5 gal (20 L) of liquid scintillation fluid and another 5-gal container equivalent of crushed glass.

flammable scintillation fluids reduces fire hazards. Those that are flammable can be incinerated or used as fuel; those that are truly biodegradable and nontoxic can be disposed of into the sewer.

Sewer Disposal

Following the NRC guidelines, aqueous wastes containing long-lived isotopes should be disposed of into the sewer. The guidelines allow for sewer disposal of up to 5 Ci/yr (185 GBq/yr) of ^{3}H, 1 Ci/yr (37 GBq/yr) of ^{14}C, and 1 Ci/yr (37 GBq/yr) of other nuclides according to 10 CFR Part 20.303 or as approved under 10 CFR Part 20.106 and 302. The tremendous dilution effected within the sewer system and wastewater treatment plant complex ensures that the radioactivity will be dispersed and not become a public health problem. However, public concerns may curtail such releases, in which case the water treatment methods used by power plants may be more widely practiced by other types of generators.

Degradation or Reclamation of Organic Liquids

As a result of a lack of coordination and agreement between the NRC and EPA on how and by whom "mixed wastes" should be regulated, at present no disposal option exists for these materials. Most mixed wastes, primarily radioactive organic liquids and oils, could be incinerated, solidified, or filtered. An alternative treatment would be to degrade them by chemical or microbiological means into water-soluble nontoxic compounds. For those that cannot be degraded, systems have been developed to separate the organic components. For example, rotary kiln furnaces are useful for treating soil or sludges, while both rotary evaporation and distillation systems can be used to separate and recover solvents (Padgett and Hollenbeck, 1989).

Gaseous Waste Treatment

Radionuclides in gaseous wastes from light-water reactors include isotopes of noble gases, activation gases, radioiodines, and tritium. Primary radionuclides from heavy-water reactors are fission products, noble gases, activation gases, radioiodines, and tritium (IAEA, 1986a). High-efficiency particulate air (HEPA) and other types of filters can be used to remove particulate matter, while charcoal and triethylenediamine (TEDA)-impregnated charcoal cartridges are useful for trapping noble gases and radioiodines. Catalytic recombination, a technology that recombines hydrogen with oxygen, can be used to reduce combustible gas mixtures. Delay systems are used to hold and decay the short-lived radionuclides in the waste stream. Longer-term delay may be provided by charcoal adsorption followed by compression of the gas into storage tanks for decay. Adsorption to charcoal, however, is a dynamic exchange process and must be viewed as only a temporary solution. Gas demisters and dryers can also be used to separate and recover mists and vapors (IAEA, 1986b).

Water Treatment

Radionuclides in the coolant system of light-water reactors exist as gases and dissolved and suspended solids. Coolant water is recirculated through a series of ion-exchange resin beds and filters to recover usable water and to meet effluent standards prior to environmental release. Various chemical processes are designed to remove the radionuclides and concentrate them for disposal. Liquid wastes are generated from the treatment of coolant water and from various subsystems within a reactor or industrial facility. These include drain wastes; ion-exchange regenerants; and wastes from decontamination operations, laundry, and miscellaneous water treatment systems (Table 3-2).

Maintaining proper water chemistry (i.e., high pH, low conductivity, and low levels of suspended solids) during active liquid waste processing can greatly extend the service life of ion-exchange resin beds and various filtration systems. The use of ion-specific resins and sluiceable demineralization systems also increases the amount of water that can be processed. One of the largest sources of liquid wastes in a power plant are demineralizer regenerants (e.g., borates or sulfates, depending upon reactor design) used to extend water treatment system life (IAEA, 1988). When system temperatures and pressures increase during power production, silica in the water can combine with cations to form a zeolite layer on fuel assemblies. This creates the potential for fuel failures, which will lead to higher-activity releases to the environment, greater potential for radioactive contamination, and increased production of radioactive waste. Silica can be removed by ion exchange, draining the tanks, and particle sedimentation (Goncarvos and Doyle, 1988). Reverse osmosis can further concentrate waste, thereby providing greater tank capacity while producing, with the aid of demineralizers, water of recyclable quality. A charcoal filter is generally placed upstream to protect both the reverse osmosis and demineralizer systems from oils in the waste (Hillmer and Swindlehurst, 1988). In combination, water treatment systems can retain up to 95% of the total dissolved solids and extend the life of demineralizers twenty-fold. Liquid wastes can be further processed with evaporators or crystallizers to increase the solids content of the wastes before solidification or volume reduction.

TABLE 3-2. Summary of LLRW by Source[a]

Types of Waste		Sources					
		Power Plants	Fuel Cycle D&D[b]	Universities	Medical	Pharmaceutical[c]	Other Industries[d]
Dry Combustible	Compactible	•	•	•	•	•	•
	Noncompactible	•	•	—	—	—	—
Noncombustible	Compactible	•	•	•	•	—	—
	Noncompactible	•	•	•	•	•	•
	Filter cartridges	•	—	—	—	—	•
Wet	Spent resins	•	•	•	—	—	—
	Sludges & slurries	•	—	—	—	—	—
	Aqueous concentrates	•	—	—	—	—	—
	Special aqueous solutions	•	•	—	—	—	—
	Oils	•	•	•	•	—	—
	Other organic liquids	•	•	—	—	•	—
	Membranes	•	—	•	•	—	—
	Biological	—	—	•	•	•	—

[a] After IAEA, 1988
[b] Decontamination and decommissioning operations
[c] Research, development, and demonstration programs
[d] Data on these wastes are incomplete and difficult to obtain

Solid Wastes

In nuclear power plants, solid wastes are generated from the operation and maintenance of the plant, effluent processing systems, and waste-volume-reduction systems. The nature of these wastes varies, but they may be classified as dry and wet wastes. "Dry" wastes include contaminated ventilation system filters, products from dryers and incinerators, contaminated plant components, charcoal, and contaminated trash such as rags, personal protective equipment, wood, paper, plastic, and tools. These are often referred to as dry active waste, or DAW. Dry wastes must contain less than 1% moisture (10 CFR Part 61.56). Dry wastes from other kinds of generators are described in Chapter 1 (Table 1-8). "Wet" waste consists of slurries such as spent ion-exchange resins, powdered resins, evaporator concentrates (or bottoms), and filter cartridges and precoat filter cakes. Wet wastes must be "dewatered" (i.e., the treated waste must contain less than 0.5% freestanding water as per 10 CFR Part 61.56) prior to disposal. Drum centrifugation, solidification, and evaporation methods are used to achieve this requirement. After dewatering, dry and formerly wet LLRW may be compressible and subjected to further pretreatment by compaction (see below).

Dewatering Systems

Sedimentation, centrifugation, filtration, compression, and forced-air drying are used to process spent resins and other wet wastes. Since criteria for freestanding water were promulgated in 1980, many methods have been established to ensure compliance. An electronic in-container moisture-detection system has been developed that is capable of determining when all interstitial liquid has been removed from a large sample volume, ensuring that less than 0.5% free liquid is in the container (Grob and Bolyard, 1988).

Linear dewatering (with and without heating) is the simplest technique and is used at most of the nuclear power plants in the United States. It uses a steel or high-density polyethylene liner that is equipped with internal perforated piping. As resin slurry is pumped into the liner, an external pump applies a vacuum and removes water via the internal piping. This process is typically a vendor-supplied service. It is an inexpensive, simple technique, but one that inefficiently utilizes the liner's inner volume in that it often leaves air pockets that prevent maximum volume reduction.

Disposal drum centrifugation produces a waste product with reduced volume (40 to 50% less than other techniques) and no free water. The inner bowl of this centrifuge is a high-integrity container. Although similar in operation to the solid-bowl centrifuge, here the drum itself becomes the disposal container, thus eliminating the need to repack and handle the processed radioactive resin (Rubin et al., 1988). (For more information on high-integrity containers, see "Conditioning Techniques" below.)

Solid-bowl centrifugation uses a continuous-feed operation that produces a dewatered resin with a low moisture content. Wastes are fed into a hollow cylinder and rotated at speeds sufficient to spin off any liquids. Unfortunately, the centrifuge is mechanically complex and frequently requires maintenance. Furthermore, as dewatered resins are removed from the centrifuge, they entrain air that can result in up to a 50% increase in volume.

Solidification involves a secondary agent that binds resin and water into a solid monolithic mixture. Containers can be of simpler construction than those used for linear dewatering and centrifugation and therefore cost less. However, this process also results in increased waste volume (see "Conditioning Techniques" below).

Compaction/Supercompaction

Since most LLRW is compressible, generators can reduce their volume of solid waste simply by compacting it. Uncompacted waste typically has a density of 113 to 159 kg/m^3. With the aid of a standard drum compactor (Figure 3-6), this can be increased to 273 to 482 kg/m^3. Shredding followed by compaction can further increase this to 642 to 722 kg/m^3. The use of more forceful horizontal piston compactors can achieve a density of 803 to 960 kg/m^3 (Ko, 1988). The new generation of compactors now on the market, called supercompactors, can yield final waste densities in the range of 960 to 1124 kg/m^3. Incineration followed by supercompaction of the resulting ash produces the greatest volume reduction. This combination yields a final density of about 11,235 to 12,840 kg/m^3—an approximately 100-fold reduction from the volume of the initial dry waste (Barbour, 1988). Some supercompactors compact entire drums of waste, as shown in Figure 3-7, that are then placed into larger overpack drums for disposal.

Incineration

Incineration can serve two purposes. It is both a volume-reduction method that can be combined with stabilization of the ash for containment in a disposal facility and a dispersal and disposal method for the volatile radioisotopes released in the effluent gases. Unlike other dis-

FIGURE 3-6. A compactor used for compressing solid LLRW. This 20,000-psi unit gives a 6:1 volume reduction for solid laboratory waste. The corrugated hose above the compactor connects the unit to a filtered exhaust line to trap particulates and gases produced during compaction.

posal methods, the releases from incinerators can be controlled and readily monitored. Incineration results in thermal breakdown of organic compounds and an average of 80:1 volume reduction (100:1 for waste of mostly papers and plastics). Of the few isotopes in LLRW that are volatile, ^3H and ^{14}C compounds are the most common and their amounts are small in comparison to the natural biospheric inventory (see Chapter 6). The ^3H and ^{14}C will escape to the atmosphere

FIGURE 3-7. (*a*) Supercompaction of whole drums of LLRW. (*b*) The effects of supercompaction on a 55-gal (209-L) drum of LLRW before and after supercompaction. These compacted drums are called "puks" and will be repackaged into a larger drum prior to shipment for disposal. (*Courtesy of Babcock and Wilcox.*)

whether the waste is buried or burned; only the rate of release will be different (IAEA, 1979; Lu and Matuszek, 1979; Matuszek, 1988). The NRC has recommended that scintillation fluids and animal carcasses containing 0.05 μCi/g (1.85 kBq/g) or less of ^{3}H and/or ^{14}C be incinerated, but has not established similar levels or recommendations for deregulation of other radioisotopes. Incineration is the method of choice for institutional and some industrial solid wastes that contain toxic chemicals, carcinogens, or pathogens. On-site incineration reduces the handling, number of people exposed, potential for transportation accidents, and the time lag between generation and final destruction of the wastes. It is also a solution for combustible mixed wastes such as organic solvents and oils.

With the exception of France, most European countries incinerate their combustible LLRW prior to storage or final disposal. Incinerator designs vary in complexity from hand-loaded models to the fully automated and heavily shielded unit used in Belgium. The latter produces a high-quality, insoluble ceramic product that further enhances waste management at the burial site. Sophisticated systems can be employed to remove particulates and hazardous gases and vapors from the effluent. The Idaho National Engineering Laboratory (INEL), for example, has designed an off-gas system that cools the air and then passes it through a HEPA filter before releasing it to the atmosphere. Scrubbing systems can further clean effluents and are especially needed when incinerating chlorinated compounds such as polyvinylchloride plastics, certain cleaning compounds, and chlorinated solvents. Incineration is more efficient and cost-effective than supercompaction, except when treating metal objects such as tools and equipment. Together, incineration and supercompaction of the resulting ash provide the highest volume reduction of any combination of methods. Concern for clean air has led to increasingly restrictive air quality standards, which make the siting, construction, and permitting of new incinerators difficult and expensive. Their utility, however, is growing as the space in landfills diminishes.

Pyrolysis

This technology is capable of providing volume reduction, dispersal of gases and vapors, and immobilization of particulates. Similar to incineration in some ways, pyrolysis employs a higher temperature and effects thermal dissociation of the waste in the absence of oxygen. Most compounds are reduced to their elemental form and discharged primarily as carbon monoxide (CO) and hydrogen (H_2). If the pyrolyzer design includes a molten glass reservoir, heavy metals from the waste can be trapped in the glass bed. Because glass has a very low leaching potential, the glass can be removed from the pyrolyzer and safely deposited in a disposal site. With a large unit, the flammable gaseous effluents (CO and H_2) can be recirculated for their fuel value. Alternatively, these effluents can be recombined with oxygen and released as carbon dioxide and water. Pyrolysis has applications similar to incineration (e.g., disposal of solid long-lived radioisotopes, pathological and toxic wastes, organic solvents, oils, spent resins), but is particularly applicable to waste materials that generate toxic chemicals upon ordinary incineration or retain radionuclides in the ash since these would be vitrified and encased as a stable, inert glass form.

Conditioning Techniques

To prevent the radioisotopes from leaching into the biosphere, a variety of immobilization techniques are available. They affect such important waste characteristics as structural stability, chemical durability, radioisotope stability, resistance against microbial attack, and gas generation. These techniques help meet the stability requirements for Class B and C wastes; some can also be adapted for use with mixed and liquid Class A wastes.

High-integrity containers (HIC) are composite, polyethylene-fiberglass reinforced plastic drums. Although HICs have been used at the Barnwell, South Carolina, site, the NRC has concluded that the polyethylene HIC does not meet its 300-year structural stability requirements (Jones, 1989). Although plastics offer excellent protection from water migration, leaching, and rusting, HICs do not have the weight-bearing strength of steel drums. Future improvements in plastics technology may increase their strength and lead to a reversal of the NRC's decision. Although cracks or holes in plastic HICs will result in water infiltration, rust will perforate most carbon steel drums in a temperate environment in less than one year (Kempf et al., 1987). The relatively rapid oxidation of steel in the presence of moisture clearly is not a reassuring standard for waste packaging. Polyethylene inner containers with larger steel overpack drums may be one interim solution that incorporates the advantages of both methods.

Cement is often used to solidify aqueous liquid wastes. It is inexpensive but also inefficient and offers only a temporary barrier to radionuclide leaching. If water infiltrates the outer container, lime may leach as a high-pH liquid and lead to accelerated decomposition of the steel drum. Cement cannot be used to solidify organic chemicals because they will readily leach out.

Bitumen and other asphalt derivatives offer more permanent matrices than cement. According to the NRC, oxidized asphalt waste qualifies for shallow land disposal in Class B and C trenches, and shows excellent resistance to leaching (WasteChem, 1988). Waste from the regeneration of ion exchange columns contains all eight heavy metals and therefore is considered both toxic and radioactive, i.e., mixed waste. Pilot projects indicate that this type of waste can be solidified in asphalt. Once solidified, it is no longer considered mixed waste and has been accepted at the present commercial disposal facilities (Hodges and Denault, 1989; Simpson, 1989; Mattus et al., 1988).

Polymers, either thermoplastics such as polyethylene, polystyrene, and polymethylmethacrylate or thermosets such as phenolics, ureas, and melamines, are suitable materials for encapsulating liquid and semi-liquid wastes. They have physical and chemical properties that make them compatible with different types of wastes and a wide range of conditions. They have a high waste-volume capacity, a low leach rate, and good chemical stability under extreme storage and disposal temperatures. Materials currently in wide use in Europe, Japan, and the United States are polyesters, epoxy resins, and styrenes (IAEA, 1988). The French have developed a process that promises to speed the curing process using thermosetting plastics.

Vitrification in borosilicate glass is the principal immobilization process used in France, Belgium, and England for the solidification of high-level radioactive waste (HLW) from fuel reprocessing. Although glass is not totally insoluble, its leaching rate is very low if water and extreme temperatures are avoided. The waste solution is first evaporated and then calcinated in a rotary kiln heated by a multi-zone furnace. The system has various off-gas treatment systems. The Atelier de Vitrification de Marcoule, France, has been in operation for only 10 years, but has over 20 years of experience in research and development of this technology (Maillet and Sombret, 1988). Although the cost of vitrification is high, use of one or a few central facilities for handling LLRW as well could reduce the overall cost by enlarging its applications to include some types of LLRW.

An emerging application of this technology is for vitrifying wastes already buried in the ground. Soil contaminated with transuranic (TRU) radionuclides at the DOE's Hanford Reservation has been vitrified in situ, resulting in a 450-ton block of glass-encapsulated soil that extends to a depth of about 7 m (Buelt and Westik, 1988) (Figure

FIGURE 3-8. (*a*) Pilot plant for in situ vitrification. The soil block to be vitrified is located below the hood where electrodes have been embedded into the ground; (*b*) results of an in situ vitrification test run, in which 20,000 pounds of soil was vitrified. (*In situ vitrification was developed at the Pacific Northwest Laboratory, which is operated for the U.S. DOE by Battelle Memorial Institute.*)

3-8). This process is very expensive as it requires large amounts of electricity. However, in situations where other methods of waste removal would create even greater contamination problems, in situ vitrification may play an important role in isolating wastes which have been improperly disposed of or begun to leach or leak.

Waste minimization and volume reduction are essential for any generator of LLRW. Reduced cost, hazard, and liability are all benefits of these actions. The available processing techniques are adequate for isolating LLRW. However, given that steel drums will be perforated by rust within one year in a trench in a temperate area, it is difficult to understand the NRC's reluctance to license high-integrity containers. In choosing any treatment method, generators should note that the additional handling associated with extended waste processing will generally increase personnel exposure to radiation.

Had volume-reduction and waste-stabilization methods been employed earlier, management at the shallow land burial (SLB) sites would have been more effective. Even the trivial but unfortunately widely publicized problems at West Valley and Maxey Flats might never have occurred, and SLB might have continued to be the method of choice for LLRW disposal.

References

Barbour, D. A. 1988. Comparative economics of the supercompaction alternative. Proceedings from the International Conference on Incineration of Hazardous/Radioactive Wastes. San Francisco, Calif.

Buelt, J. L., and J. H. Westik. 1988. In situ vitrification—preliminary results from the first large-scale radioactive test. Waste Management '88 1:711–723.

Goncarvos, B., and R. D. Doyle. 1988. A cost effective approach to silica reduction. Waste Management '88 1:483–486.

Grob, L. H., and Bolyard, F. K., Jr. 1988. End-point moisture detection for dewatered media—a new approach. Waste Management '88 1:487–491.

Hillmer, T. P., and D. P. Swindlehurst. 1988. Use of a reverse osmosis system for testing radwaste at Palos Verde. Presented at Waste Management '88. Tucson, Ariz.

Hodges, S. M., Jr., and R. P. Denault. 1989. Emergency avoidance solidification campaign of liquid low-level waste at the Melton Valley storage tank facility. Waste Management '89 2:751–755.

International Atomic Energy Agency (IAEA). 1979. Behaviour of Tritium in the Environment. Vienna, Austria: IAEA.

International Atomic Energy Agency (IAEA). 1986a. Design of Radioactive

Waste Management Systems at Nuclear Power Plants. Safety Series No. 79. Vienna, Austria: IAEA.

International Atomic Energy Agency (IAEA). 1986b. *Operational Management for Radioactive Effluents and Wastes Arising in Nuclear Power Plants.* Safety Series No. 50-SG-011. Vienna, Austria: IAEA.

International Atomic Energy Agency (IAEA). 1988. *Immobilization of Low and Intermediate Level Radioactive Wastes with Polymers.* Technical Report No. 289. Vienna, Austria: IAEA.

Jones, D. 1989. 1988—Year of the high integrity container—evaluation, controversy and regulatory actions. *Waste Management '89* 2:389–394.

Kempf, C. R., C. J. Sven, S. Mughabghab, and T. M. Sullivan. 1987. *Low-Level Waste Source Term Evaluation: Quarterly Progress Report, January-March 1987.* WM-3276-2. Upton, N.Y.: Brookhaven National Laboratory.

Ko, S. 1988. Analysis of the reduction in waste volumes received for disposal at the low-level radioactive waste site in the state of Washington. *Waste Management '88* 1:629–635.

Lu, A. H., and J. M. Matuszek. 1979. Transport through a trench cover of gaseous tritiated compounds from buried radioactive wastes. In *Behaviour of Tritium in the Environment.* Vienna, Austria: IAEA.

Maillet, J., and C. Sombret. 1988. High-level waste vitrification: The state of the art in France. *Waste Management '88* 2:165–172.

Mattus, A. J., R. D. Doyle, and D. P. Swindlehurst. 1988. Asphalt solidification of mixed wastes. *Waste Management '89* 1:229–234.

Matuszek, J. M. 1988. Safer than sleeping with your spouse—the West Valley experience. In *Low-Level Radioactive Waste Regulation: Science, Politics, and Fear,* ed. Michael E. Burns, Chelsea, Mich.: Lewis Publishers, pp. 261–277.

National Low-Level Radioactive Waste Management Program (NLLRWMP). 1987. *LLRW Management in Medical and Biomedical Research Institutions.* DOE/LLW-13th. Washington, D.C.: U.S. Government Printing Office.

Padgett, D., and P. Hollenbeck. 1989. X*TRAX™: An application for mixed waste separation. *Waste Management '89* 2:519–523.

Party, E. P., and E. L. Gershey. 1989. Recommendations for radioactive waste reduction in biomedical/academic institutions. *Health Physics* 56(4):571–572.

Rubin, L. S., C. P. Deltete, and M. R. Crook. 1988. Radwaste disposal drum centrifuge. *Waste Management '88* 1:327–334.

Simpson, S. 1989. Waste form testing results for DOE and commercial power reactor wastes. *Waste Management '89* 2:301–304.

Spall, M. J., M. L. Miele, and W. A. Homyk. 1988. Reduction at Indian Point Units 1 & 2: Turning around a large volume generator. *Waste Management '88* 1:403–409.

WasteChem Corporation. 1988. U.S. NRC approves WasteChem's VRS™ system waste form: topical Report. *News from WasteChem.* Paramus, N.J.: WasteChem.

4 DISPOSAL METHODS

The objective of any waste disposal strategy is to prevent interaction between the waste and the environment. Depending upon the characteristics of, and the hazards presented by, the waste, different treatment methods exist to contain, disperse, or render it nonhazardous. For instance, destruction by high-temperature incineration or chemical inactivation is appropriate for most hazardous chemicals and infectious wastes. For domestic sewage, biological degradation in a properly operated wastewater treatment plant is sufficient to make the effluent harmless. Unlike other types of hazardous waste, the hazard associated with radioactive waste, its radioactivity, cannot be eliminated or otherwise destroyed, but it will decrease with time. However, the hazard from radiation will persist as long as the radioisotope is decaying and emitting energy and no method exists to reduce or stop radioactive decay. Thus, disposal methods for radioactive wastes must rely on either dilution and dispersal techniques for low-hazard materials or isolation and confinement for high-hazard wastes. The level of sophistication of the selected disposal method should be directly related to the hazard presented by the waste. Selection of inappropriate, overly conservative disposal methods adds unnecessary expense and increases risk to workers and/or the public (IAEA, 1985).

In the past, the United States, as well as most other nations that produce radioactive waste, used the oceans as disposal grounds for low-level radioactive waste (LLRW). From 1946 to 1970, some 201,500 drums of about 80-gal capacity each were disposed of into the Atlantic and Pacific oceans (Carter and Stone, 1986). These drums, primarily DOE wastes, contained a radionuclide inventory of about 214,300 Ci (7929 TBq). In 1960 the Atomic Energy Commission placed a moratorium on the issuance of new licenses for sea disposal, and since then most LLRW has been disposed of at shallow land burial (SLB) facilities. At present "The Commission will not approve any application for a license for disposal of licensed materials at sea unless the applicant shows that sea disposal offers less harm to man or the environment than other practical alternative methods of disposal" [10 CFR Part 20.302(b)]. The proposed revision to 10 CFR Part 20 would delete this section, reflecting the mandate of the 1972 Marine Protection, Research, and Sanctuary Act, which transferred responsibility for regulating the ocean disposal of radioactive wastes from the Nuclear Regulatory Commission (NRC) to the Environmental Protection Agency (EPA). Ocean disposal was used on a regular basis by Belgium, Netherlands, Switzerland, Japan, and the United Kingdom, and occasionally by France, Germany, Italy, and Sweden. Although scientific evidence cannot be found to show that this practice did environmental damage (U.S. GAO, 1984), ocean disposal was stopped because of increased opposition by environmental activist groups. Ocean disposal has gained some renewed interest as a possible method for high-level radioactive waste (HLW) disposal (Gomez, 1986). Improvements over simple dumping such as subseabed emplacement of waste containers could make ocean disposal a more attractive and environmentally sound procedure, but it is unlikely, given the political pressures, that any country will revert to ocean dumping for LLRW.

Disposal Strategies Permitted by the NRC

Currently the NRC permits the disposal of LLRW by the following six methods under 10 CFR Part 20. Only the first one, the transfer and disposal of LLRW, will be discussed in depth here. The other methods have been discussed in Chapter 3. These NRC-approved disposal methods are:

1. *Transfer to an authorized recipient* [10 CFR Part 20.301(a)]. The waste may be physically moved from the point of generation to an NRC-approved transporter, treatment, or disposal facility. Although the waste is transferred, title to and liability for the waste remains that of the generator.
2. *Special ruling* [10 CFR Part 20.301(b) and 302]. By petitioning the NRC or its agreement state agency for approval of an alternative method of treatment for a unique portion of a LLRW waste stream, such as on-site decay or deregulation, the generator may dispose of that portion through means not necessarily listed under 10 CFR Part 20.

3. *Release into sanitary sewerage systems* [10 CFR Part 20.301(c) and 303]. A defined amount of radioactivity, in aqueous liquid form, may be disposed of into a sanitary sewer system.
4. *Release of air and liquid effluents* (10 CFR Part 20.106). Given that the material does not exceed a defined concentration limit (maximum permissible concentration or MPC), generators are allowed to release radioactivity into air or water systems. This method is primarily used by power plants for the release of slightly contaminated steam and cooling water.
5. *Disposal of specific wastes without regard to radioactivity* (10 CFR Part 20.306). In this case, the specific wastes are liquid scintillation fluids and animal tissues containing not more than 0.05 μCi (1.85 kBq) of ^{3}H or ^{14}C per gram. The NRC has been asked to expand the list of wastes that can be disposed of under this regulation.
6. *Incineration.* This is the method recommended by the NRC (10 CFR Part 20.306) for the disposal of deregulated scintillation fluids and animal tissues. Specific approval must be sought for other types of LLRW.

Disposal Technologies

Although there is no universal method for disposing of LLRW, most countries do use some form of land repository. Land repositories range from shallow land burial (SLB) sites, such as the trench facilities used in the United States since the early 1960s, to deep geological repositories designed for long-term containment of the wastes, such as the SFR facility mined in crystalline rock in Sweden. Design of a facility is determined by the kind of waste to be disposed of and the feasibility of the option chosen (IAEA, 1981b). In the United States, facilities must be sited, designed, operated, closed, and subsequently controlled to meet the performance objectives of 10 CFR Part 61. Releases to the environment must be kept as low as reasonably achievable (ALARA) and waste confinement should remain effective until the radionuclides have decayed to acceptable levels before entering the environment. These objectives also state that the general population must be protected from releases of radioactivity, individuals must be protected during operations at the site, the site must be protected against inadvertent intrusion, and site stability must be ensured after closure. The site must remain intact for at least 300 years after closure. Concentrations of radioactive materials from the facility that might be released to the general environment in ground and surface water, air, soil, plants, or animals cannot exceed an annual dose to any member of the general population of 25 mrem (0.25 mSv) to the whole body, 75 mrem (0.75 mSv) to the thyroid, or 25 mrem (0.25 mSv) to any other organ (dose is discussed in Chapter 6). The highest dose measured near a repository is 1 mrem (0.01 mSv) (Eisenbud, 1987). Given the current criteria for disposal facilities, the projected radiation dose from an LLRW facility is 0.003 mrem/yr (0.00003 mSv/yr) (Eisenbud, 1988). Exposure to radiation from an LLRW disposal site can be limited by several means, for example, delay release of radionuclides, retard their transport or dilute their concentration to assure that the impact on humans will be within acceptable levels (IAEA, 1981b). Site selection must be made carefully and exhaustively and should account for at least the following four factors: hydrogeology, ecology, land use, and socioeconomics (IAEA, 1981a).

A disposal facility consists of a plot of land within which the disposal site and support buildings are constructed. The actual disposal site consists of a disposal unit, such as a trench and an encircling buffer zone. The buffer zone provides a controlled space in which to establish monitoring locations that provide early warnings of any radionuclide migration. Land disposal methods, listed below and compared in Table 4-1, are of two types: shallow ground disposal, where disposal is made within 30 m of the earth's surface or above the surface, and deeper disposal, as in mined cavities of natural or manmade origin. Sanitary landfills are not covered by 10 CFR Part 61 (see Chapter 2).

Shallow Land Burial (SLB)

In a typical SLB facility, wastes are buried within 30 m of the earth's surface and later covered with soil and other overburden materials (Figure 4-1) (10 CFR Part 61). The disposal unit is usually a trench, which may or may not be lined with materials such as concrete or heavy-gauge plastic sheeting. Waste is emplaced using cement casks, steel drums, or bins as containers. The trenches, once full, are backfilled with the excavated earth, compacted and capped, usually with a clay layer to minimize water infiltration. A gently sloped mound of soil placed over this impermeable layer provides further protection against water infiltration by providing a surface for runoff. Finally, the surface of the trench is planted with vegetation for stabilization. Various methods to control water accumulation within the trenches

TABLE 4-1. Comparison of Land Disposal Facilities[a]

Site Features	SLB	BGV	AGV	EMCB	SD	MC
Operational experiences	Used as the commercial option since 1962	As retrievable storage in US & Canada	Short-term storage in Canada	France since 1960s, in Canada & U.S. (ERDA)	Under study by DOE, used by Los Alamos	Used in Germany (Asse mine) and Sweden (SFR)
Suitability for long-term isolation of wastes relies on	Site characteristics	Site characteristics	Design features of vaults; has more siting flexibility	Site characteristics and design of structures	Site characteristics	Site characteristics, structural geology is critical
Select area free of flooding & ponding	Yes	Yes	Yes	Yes	Yes	N/A
Susceptible to groundwater intrusion	Yes	Only if vault's integrity is compromised	No	Yes, especially upper section	Yes	Yes, a serious threat
Need for protection against precipitation during site operation	Yes	Yes	No	Yes	Yes	No
Need for geochemical compatibility of soil and engineered structure	No	Yes	No	Yes	Yes if lining used	No
Appropriateness of site design features for long-term isolation	Adequate	Adequate	Relies on single engineered barrier; subject to natural disasters	Adequate	Adequate	Adequate

(Table 4-1 continues)

TABLE 4-1. Comparison of Land Disposal Facilities[a] (cont'd)

Site Features	SLB	BGV	AGV	EMCB	SD	MC
Susceptibility to natural disasters	Low	Low	High	Low	Low	Moderate
Long-term maintenance requirements after closure	Low	Low	Frequent & expensive	Low	Low	Low
Buffer zone & monitoring system must provide early warning of radionuclide releases	Adequate; soil around trench retards radionuclide migration to surface & groundwater	Adequate; soil around vault retards radionuclide migration to surface & groundwater	If monitoring at vault's surface, no mitigative measures possible before unacceptable releases might occur	Adequate; soil around site retards radionuclide migration to surface & groundwater	Adequate	Adequate
Need for filling voids to minimize site subsidence after closure. In all of them it retards radionuclide migration	Yes	No	No	Yes	Yes	No
Need of segregation of unstable Class A waste for structural support, see general comments	Essential to avoid active maintenance after closure	N/A, but waste decomposition products could chemically damage vault structure	N/A, but waste decomposition products could chemically damage vault structure	Essential	Essential	N/A, but waste decomposition may build up explosive gases

Fire hazard associated with combustible waste & methane buildup	Low	Low	High, incineration essential	Low	High	High, incineration essential
Amenable to use of remote handling equipment	Yes	No	No	Yes	Yes	Yes
Worker exposure in mrem/yr[d]	245 avg at Barnwell, S.C., 1985[b].	>SLB; use of remote handling would reduce it, but is hampered by the limited access	>SLB; use of remote handling would reduce it, but is hampered by the limited access	1000 avg 2000 max at La Manche, France, 1981[b]	Equal or less than SLB depending on procedure	Waste handling: 1500; background: 500 for salt & clay, 1500 for hard rock mines[b]
Worker dose equivalent estimated by Herrington et al. in person-rem/m³ [c,e]	---	1.47×10^{-3}	1.6×10^{-3}	1.81×10^{-3} [f]	1.29×10^{-3} for augered hole	2.06×10^{-3}
Preoperational, capital, operational, closure & postclosure relative expenses	$	$$	$$$	$$$$	$$$$$	$$$$$$

SLB: shallow land burial; BGV: below-ground vault; AGV: above-ground vault; EMCB: earth-mounded concrete bunker; SD: shaft disposal; MC: mine cavities; N/A: not applicable.

[a] Bennett, 1985; Bennett et al., 1985; Bennett and Warriner, 1985; McAneny, 1986; Miller and Bennett, 1985; Warriner and Bennett, 1985

[b] NYS DEC 1987

[c] Herrington 1987

[d] 1 mrem = 0.01 mSv

[e] 1 person-rem = 0.01 person-Sievert

[f] Note that the estimate for EMCB does not agree with measured values at La Manche.

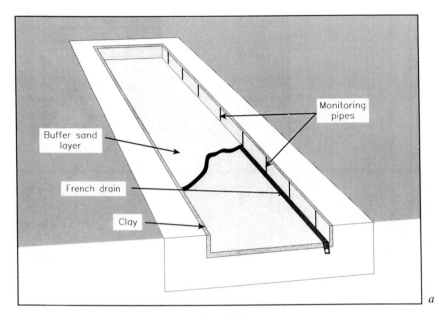

Monitoring
pipes

Buffer sand
layer

French drain

Clay

a

Ground level

Topsoil

Compacted clay layer

Compacted sand

Monitoring
pipe

Waste packages

French drain

Buffer sand

b

104

FIGURE 4-1. Shallow land burial (SLB) disposal facility. (*a*) Trenches are approximately 100 ft wide, 1000 ft long, and 20 to 30 ft deep, and include a subsurface drainage system and monitoring pipes; (*b*) cross-section of a disposal trench; (*c*) aerial view of trench at the Barnwell, S.C., facility; (*d*) waste packages emplaced at Barnwell. (*Photographs courtesy of Chem-Nuclear Systems, Inc.*)

are implemented, using a combination of sloping trench floors, sumps, and riser pipes for the observation of water levels or water removal (IAEA, 1984; Dayal et al., 1986). SLB has been used at all six of the commercial U.S. LLRW disposal sites as well as in other countries, for example, France and the U.S.S.R., for nearly three decades (IAEA, 1985). Water management problems that led to measurable but trivial releases of radionuclides have occurred at certain facilities. Most of these can be traced back to improper site selection, the nature of the waste, or poor packaging and disposal practices. It is important to note that little guidance was available from the NRC in the early days of SLB practices and waste packaging criteria were not nearly as stringent as they are today under 10 CFR Part 61 and DOT regulations. Nevertheless, experience shows that many of these problems could have been avoided by the use of appropriate waste packaging and siting criteria as well as more careful management of the disposal facility (IAEA, 1985).

Below-Ground Vault (BGV)

BGVs are enclosed, engineered structures built totally below the surface of the earth (Figure 4-2). They can be constructed of masonry blocks, fabricated metal shapes, reinforced cast-in-place or sprayed concrete, precast concrete, or plastic or fluid media molded into various solid shells. The vault can be built with engineered walls and roof; the floor can be natural soil or rock, treated soil or rock, or engineered materials. The vault has limited access to its interior space through a doorway or portal. Operational access to the vault from the surface may be in the form of an excavated ramp that is covered at the time of closure (10 CFR Part 61). Oak Ridge National Laboratory uses BGVs for the retrievable storage of transuranic (TRU) waste in their Solid Waste Storage Area No. 5. Shallow BGVs are also used in Canada at the Chalk River Nuclear Laboratory and Whiteshell Nuclear Research Establishment (Warriner and Bennett, 1985).

Above-Ground Vault (AGV)

An AGV is an engineered structure or building with floors, walls, roof, and limited access openings on a foundation near the ground surface (Figure 4-3a). At least some portion of the structure is above the postclosure surface grade. Fabrication can be of masonry blocks, fabricated metal shapes, reinforced cast-in-place or sprayed concrete,

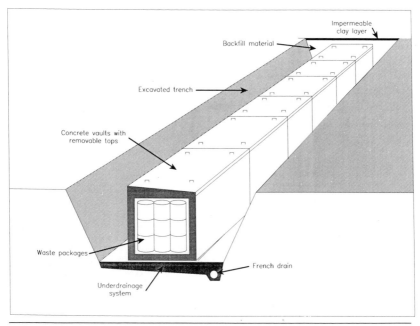

FIGURE 4-2. Below-ground vault (BGV) disposal facility. Compartmental-ized concrete vaults are placed in an excavated trench that has a drainage system. Vaults have removable roofs. After individual vaults are filled and the roof emplaced, the vault is covered with permeable material and then the trench is covered with an impermeable clay layer and topsoil and seeded with vegetation. *(After Warriner and Bennett, 1986.)*

precast concrete, or plastic or fluid media molded into various solid shells (10 CFR Part 61). There is only short-term storage experience in the United States with AGVs [an AGV has been in use at the West Valley Demonstration Project in New York since 1986 (Figure 4-3b)], but in Canada this technique has been used for storage of LLRW for approximately 10 years with AGVs at both the New Brunswick Elec-tric Power Commission's Point Lepreau site and Ontario Hydro's Bruce site (Bennett and Warriner, 1985).

Earth-Mounded Concrete Bunker (EMCB) or Tumuli

EMCBs include a combination of the features used in other meth-ods, for example, trenches, BGVs, and earthen mounds (tumuli) (Fig-ure 4-4). Site integrity relies upon waste form packaging and site operation to ensure control of radionuclide release. France has been

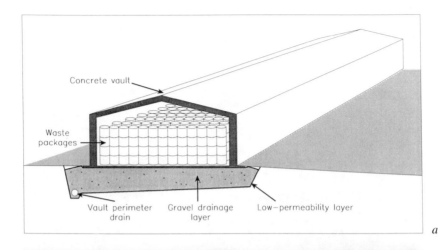

FIGURE 4-3. Above-ground vault (AGV) disposal facility. (*a*) Cross section shows foundation with gravel drainage layer, drainage system, and peripheral subsurface drain *(after Bennett and Warriner 1986)*; (*b*) AGV with wastes in place at the West Valley Demonstration Project in New York. Square drums are used to reduce wasted space and facilitate stacking. *(Courtesy of West Valley Demonstration Project.)*

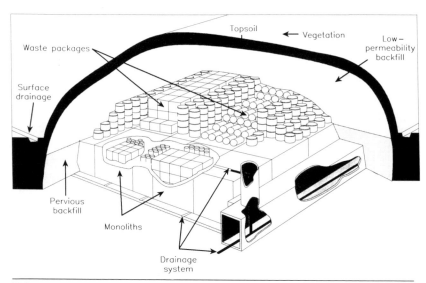

FIGURE 4-4. Earth-mounded concrete bunker (EMCB) disposal facility. Wastes are separated according to activity—higher-activity wastes are placed in the below-ground monoliths, lower-activity wastes are placed on top of the monoliths to create the mounds. The bunker is equipped with a drainage system within and around the structure. The wastes are covered with permeable material, then an impermeable clay layer and topsoil and seeded with vegetation. *(After Miller and Bennett, 1986.)*

using this technology since 1969 at Centre de La Manche and will open a new facility in 1990 at Centre de l'Aube. The bunkers currently used in France segregate waste according to the level of radioactivity. Wastes with lower levels are placed above ground at natural grade in tumuli, while wastes with higher levels of radioactivity are embedded in concrete monoliths below ground. A variation of the EMCB technology was tested by the Energy Research and Development Administration (ERDA) for the disposal of TRU solid waste. Another variation has been used in Canada for the storage of LLRW at the Chalk River National Laboratory and the Whiteshell Nuclear Research Establishment (Miller and Bennett, 1985). The DOE has a demonstration program using the tumulus technology (Garland et al., 1989). Several variations on this technology have been proposed in the United States, such as earth-covered concrete overpacks (concrete overpacks containing waste are stacked over a concrete pad and then covered with soil) and earth-covered waste modules (waste packages, without concrete overpacks, are stacked from above into a concrete waste module) and are being evaluated (Conner et al., 1988; Eng et al., 1989; Liu, 1989; Marschke and Anigstein, 1989).

Shaft Disposal (SD)

SD is the use of shafts and boreholes augered, bored, or sunk into the ground by conventional methods (Figure 4-5). The shafts can be lined or unlined and of various sizes. Lining can be made of concrete, metal, or other suitable low permeability materials (10 CFR Part 61). The DOE, at its Nevada Test Site, is evaluating the use of large-diameter augered holes for the disposal of high-specific-activity LLRW. Actual waste emplacement began in 1983. As part of the same program, the DOE plans to dispose of Class B waste by SD at the Savannah River Plant. At Oak Ridge National Laboratory, TRU waste is being stored in shallow holes, and at Los Alamos National Laboratory, augered holes have been used for the disposal of solid wastes that require external shielding. Tritium and TRU waste were first encased in asphalt in 55-gal (209-L) drums before their emplacement in the shafts. In Canada, concrete pipes on concrete foundations have been used for the storage of ion-exchange resins and filter canisters at Ontario Hydro's Bruce site and the Chalk River Nuclear Laboratory (Bennett, 1985).

Rock Cavities

Solid or solidified waste can be emplaced in existing man-made or natural cavities such as mines, or in cavities specially excavated for this purpose (Figure 4-6). These cavities may be sited in different geological formations and located at different depths, from near the surface to depths typical for deep geological disposal. The stability of such disposal facilities will be dictated by the integrity and features of the local geology. Although intact rock formations offer excellent containment and low hydraulic permeability, such formations, if fractured, may lead to rapid migration of radionuclides. Rock cavity repositories exist in Czechoslovakia, Federal Republic of Germany (FRG), German Democratic Republic, and Spain. The FRG disposed of waste in an abandoned salt mine from 1967 to 1978. Several countries are in the process of developing rock cavity facilities (IAEA, 1985). In the United States, the DOE is developing a deep repository for TRU waste in a rock salt formation at the Waste Isolation Pilot Project (WIPP) near Carlsbad, New Mexico. The federal government is also pursuing the development of mined repositories for the disposal

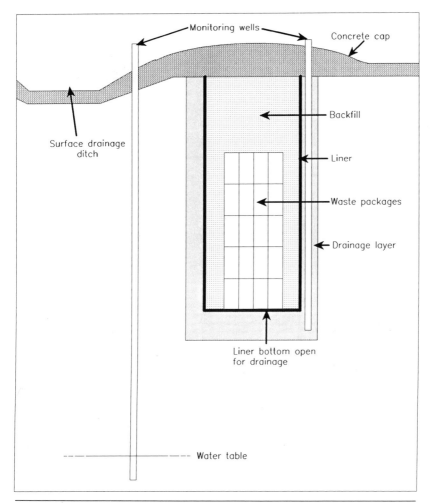

FIGURE 4-5. Shaft disposal (SD) facility designed for disposal of wastes above the groundwater table. The shaft has both in-shaft and surface drainage systems and monitoring wells for the shaft and the surrounding area. *(After Bennett, 1985.)*

of HLW at Yucca Mountain in Nevada. The use of active and abandoned mines for LLRW is currently under evaluation by the NRC (10 CFR Part 61).

The only commercial experience in the United States with the land disposal of LLRW has been SLB. However, there is in-field experi-

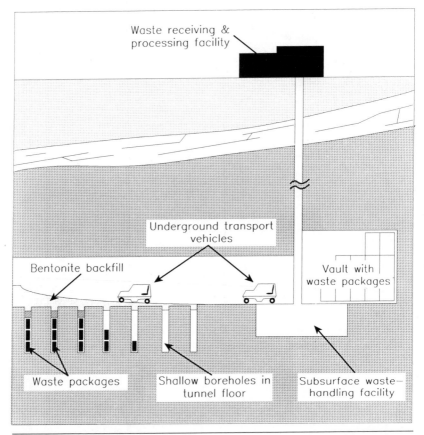

FIGURE 4-6. Rock cavity disposal facility. Wastes are placed in vaults or in shallow boreholes within a mined cavity. The cavity itself may be as deep as 3000 m. Although a facility would usually employ either boreholes or vaults, both are shown here. The cavity and vaults may be backfilled with a buffer material such as bentonite. *(After Chapman and McKinley, 1987; Murray, 1989.)*

ence with all of the methods, and all of the disposal strategies described above, either alone or in combination, are adequate for the disposal of LLRW. The characteristics and disadvantages of the different strategies are listed in Table 4-1. For a strategy to be adequate, it must ensure that no member of the public will be irradiated by the radioactivity in the LLRW, that the waste will be contained for long periods of time, and that any migration of radionuclides through soil or groundwater from the site will be in amounts too small to pose health risks. SLB is the least costly method. As the design becomes increasingly "engineered," costs increase. The mined cavity design is the most expensive and therefore has been used primarily for HLW that requires long-term isolation. The EMCB and mined cavities lead to much higher worker doses than those measured at SLB facilities. The AGV is different from the others because it is not a "burial" method. Therefore its siting needs are fairly independent of the local geology and weather. However, it has no soil buffer zone to retard radionuclide migration, requires more extensive and expensive maintenance after closure, is more susceptible to intrusion, and results in higher worker doses than SLB. It is also more subject to damage from natural disasters such as earthquakes and tornadoes. An important point to remember is that whereas the behavior of geological formations over the next 1000 years can be predicted, we lack the data and experience to make the same predictions about concrete (Bennett and Warriner, 1985; MacKenzie et al., 1986).

Because the success of all of these methods during both operation and postoperational phases depends upon waste management, strict rules on the preparation and placement of wastes are needed. The types and amounts of radionuclides present, the nonradioactive hazards of the waste, and the package's ability to withstand not only overburden pressure but also chemical and biological degradation will all have an effect on the success of the disposal strategy (IAEA, 1985). Separate disposal areas within the disposal facilities should be designed and developed based upon the concentration and half-life of the materials that they will receive. Special attention must be given to nuclides like ^3H that have a relatively long half-life and the potential for migration as liquid or vapor. To enable better site management, isotopes with half-lives longer than 100 years should be disposed of in separate areas. Short-lived (i.e., <90-day half-life wastes) and deregulated wastes should be excluded from disposal at radioactive waste sites, eliminating a major fraction of the Class A waste. Since Class A waste does not have the same stability requirements as Classes B and C, after disposal it may decompose, releasing gases and liquids that could damage the disposal site structure and/or present a fire hazard (Siskind et al., 1985). For these reasons, unstable Class A wastes that must be buried should first be treated by incineration or one of the conditioning techniques described in Chapter 3. Wastes containing additional nonradiological hazards should be treated by incineration, pyrolysis, or vitrification to eliminate these hazards before disposal. Unlike Class A waste, the stability requirements for Class B and C wastes can be achieved by the waste form itself or by a disposal container or structure that provides stability after disposal. Structurally stable waste forms must endure overburden weight, compaction, presence of moisture, microbial activity, and internal factors such as radiation effects and chemical changes. However, as described in Chapter 2, water management has been the single most important factor in determining the success of the existing facilities. Minimizing the access of water reduces the potential for radionuclide migration, eliminates the need for long-term active maintenance, and reduces potential exposures to the public.

References

Bennett, R. D., W. O. Miller, J. B. Warriner, P. G. Malone, and C. C. McAneny. 1984. *Alternative Methods for Disposal of Low-Level Radioactive Wastes. Task 1: Description of Methods and Assessment of Criteria.* NUREG/CR-3774, Vol. 1. Washington, D.C.: NRC.

Bennett, R. D. 1985. *Alternative Methods for Disposal of Low-Level Radioactive Wastes. Task 2e: Technical Requirements for Shaft Disposal of Low-Level Radioactive Waste.* NUREG/CR-3774, Vol. 5. Washington, D.C.: NRC.

Bennett, R. D., and J. B. Warriner. 1985. *Alternative Methods for Disposal of Low-Level Radioactive Wastes, Task 2b: Technical Requirements for Aboveground Vault Disposal of Low-Level Radioactive Waste.* NUREG/CR-3774, Vol. 3. Washington, D.C.: NRC.

Carter, M. W., and D. C. Stone. 1986. Quantities and Sources of Radioactive Waste. In *Radioactive Waste,* NCRP Proceedings No. 7, Bethesda, Md.: NCRP, p. 5-30.

Chapman, N. A., and I. G. McKinley. 1987. *The Geological Disposal of Nuclear Waste.* Chichester, Great Britain: John Wiley & Sons.

Conner, J. E., J. T. Case, V. C. Rogers, and R. Eng. 1988. Low-level radioactive waste prototype license application project. *Waste Management '88* 1:265–267.

Dayal, R., R. F. Pietrak, and J. H. Clinton. 1986. *Geochemical Studies of Commercial Low-Level Radioactive Waste Disposal Sites.* NUREG/CR-4644. Washington, D.C.: NRC.

Eisenbud, M. E. 1987. *Environmental Radioactivity From Natural, Industrial, and Military Sources,* 3d ed. Orlando, Fla.: Academic Press.

Eisenbud, M. E. 1988. Low-level radioactive waste repositories—a safety assessment. *Annual Conference of the North Carolina Academy of Sciences.* 26 March 1988, Charlotte, N.C.

Eng, R., P. Liu, W. Chang, and I. Tsang. 1989. Earth mounded concrete bunker disposal system. *Waste Management '89* 2:101–106.

Garland, S., S. D. Van Hoesen, T. F. Scanlan. 1989. Experience with the tumulus technology for the disposal of solid LLRW. *Waste Management '89* 2:23–28.

Gomez, L. S. 1986. Subseabed disposal of high-level radioactive waste. In *Radioactive Waste*, NCRP Proceedings No. 7. Bethesda, Md.: NCRP, pp. 122–132.

Herrington, W. N., R. Hartz, S. E. Merwin. 1987. *Occupational Radiation Exposures Associated with Alternative Methods of Low-Level Waste Disposal*. NUREG/CR-4938. Washington, D.C.: NRC.

International Atomic Energy Agency (IAEA). 1981a. *Shallow Ground Disposal of Radioactive Wastes—A Guidebook*. Safety Series No. 53. Vienna, Austria: IAEA.

International Atomic Energy Agency (IAEA). 1981b. *Underground Disposal of Radioactive Wastes—Basic Guidance*. Safety Series No. 54. Vienna, Austria: IAEA.

International Atomic Energy Agency (IAEA). 1984. *Safety Analysis Methodologies for Radioactive Waste Repositories in Shallow Ground. Procedures and Data*. Safety Series No. 64. Vienna, Austria: IAEA.

International Atomic Energy Agency (IAEA). 1985. *Acceptance Criteria for Disposal of Radioactive Wastes in Shallow Ground and Rock Cavities*. Safety Series No. 71. Vienna, Austria: IAEA.

Liu, P. 1989. Stability analysis of the earth mounded concrete bunker disposal system. *Waste Management '89* 2:771–774.

MacKenzie, D. R., B. Siskind, B. S. Bowerman, and P. L. Piciulo. 1986. *Preliminary Assessment of the Performance of Concrete as a Structural Material for Alternate Low-Level Radioactive Waste Disposal Technologies*. NUREG/CR-4714. Upton, N.Y.: Brookhaven National Laboratory.

Marschke, S., and R. Anigstein. 1989. Safety assessment of the earth mounded concrete bunker disposal system. *Waste Management '89* 2:761–764.

McAneny, C. C. 1986. *Alternative Methods for Disposal of Low-Level Radioactive Wastes: Task 2d: Technical Requirements for Mined Cavity Disposal of Low-Level Radioactive Waste*. NUREG/CR-3774. Washington, D.C.: NRC.

Miller, W. O., and R. D. Bennett. 1985. *Alternative Methods for Disposal of Low-Level Radioactive Wastes, Task 2c: Technical Requirements for Earth Mounded Concrete Bunker Disposal of Low-Level Radioactive Waste*. NUREG/CR-3773, Vol. 4. Washington, D.C.: NRC.

Murray, R. L. 1989. *Understanding Radioactive Waste*. 3d ed. Columbus, Ohio: Battelle.

New York State Department of Environmental Conservation (NYSDEC). 1987. *Final Environmental Impact Statement for: Promulgation of 6 NYCRR Part 382: Regulations for Low-Level Radioactive Waste Disposal Facilities*, Vols. I and II. New York: NYSDEC.

Siskind, B., D. R. Dougharty, and D. R. MacKenzie. 1985. *Extended Storage of Low-Level Radioactive Waste: Potential Problem Areas*. NUREG/CR-4062. Washington, D.C.: NRC.

U.S. General Accounting Office (U.S. GAO). 1984. *Hazards of Past Low-*

Level Radioactive Waste Ocean Dumping Have Been Overemphasized.
Report EMD-82-9. Washington, D.C.: U.S. Government Printing Office.
Warriner, J. B., and R. D. Bennett. 1985. *Alternative Methods for Disposal of Low-Level Radioactive Wastes, Task 2a: Technical Requirements for Belowground Vault Disposal of Low-Level Radioactive Waste.* NUREG/CR-3774, Vol. 2. Washington, D.C.: NRC.

5 STATUS OF STATE PLANS AND REGIONAL COMPACTS

The federal government recognized in 1979 that the disposal capacity of the existing sites would be inadequate for future needs. Instead of instituting regulations for volume reduction and waste minimization and planning for a central federal site for the nation's low-level radioactive waste (LLRW) as a whole, as a number of European countries have done, the Congress passed legislation making each state responsible for its own LLRW. Why the United States is following this congressionally defined route to a solution for the problems of LLRW disposal is complex. In the late 1970s and early 1980s, the federal government, in an effort to reduce its size, shed itself of numerous responsibilities, handing them over to the state governments. During this period, the governors of the states with disposal sites (South Carolina, Nevada, and Washington) recognized the diminishing capacity of their sites and in 1979 closed their sites, threatening permanent closure if the federal government did not make provisions for future LLRW disposal. In such a political climate, the Low-Level Radioactive Waste Policy Act of 1980 was passed with little ado, and the responsibility for LLRW management was handed over to the states. The transition has been far from smooth. The three states with existing disposal sites quickly formed regional compacts with other states. The remaining states had little incentive to meet the first set of deadlines. Congressional representatives who were not from the states petitioning for compact status slowed action on ratification. Consequently, none of the states met the first deadline. Congress extended these deadlines and added penalties for inaction with the passage of the Low-Level Radioactive Waste Policy Amendments Act (LLRWPAA) of 1985.

A number of questions have arisen since the passage of the Policy Act, with answers still forthcoming. Apart from the question of whether the federal government has the ability to order states to assume federal responsibilities and enforce federal regulations, the compacts are inadequate to protect the host state. For example, how binding is a compact arrangement which must be reevaluated every 5 years, when the possibility exists that congressional approval may be withdrawn after the reevaluation? Do the compacts have the right to impose regulations on the host state that are in conflict with those of the host state? What if the definition of LLRW is changed, resulting in an increase in radioactive concentration that the site is not equipped to accept? Conversely, volume-reduction measures and deregulation would adversely affect the large volume of waste crucial to support the site economically. Will states and compacts have the ability to exclude the import or restrict the export of wastes despite the supremacy and commerce clauses? What will happen when major generators go out of business? Will the other states or generators cover the financial shortfall? What will guarantee that host states will spend the compact money wisely? Will the compacts assume responsibility for the site, its insurance, and the liability of the waste or will that fall upon the host state? Can the states or companies actually take title to the waste despite the RCRA and Superfund (CERCLA) rules that define the generator's liability as "cradle to grave"? If states continue along this route and compacts develop operating sites, it is uncertain what the relationship between compacts will be or how centralized treatment centers will be accessed. Considering that the deadlines are just around the corner, these and other questions cast serious doubt on the effectiveness and enforceability of the Amendments Act, and the ability of the states to comply with it.

In forcing each state to provide for the disposal of its own LLRW, the Amendments Act may be creating a situation in which the generators cannot afford to dispose of their wastes at the new sites. Economic, regulatory, technical, and procedural problems have surfaced during the effort to meet the deadlines. As of early 1990, seven unaligned states and nine compacts (Figure 5-1) were reinventing the regulatory and developmental wheel, expending tax dollars to establish agencies and committees that will select, characterize, and designate sites and disposal technologies. By the end of 1989, the Illinois Department of Nuclear Safety and the Central Midwest Compact spent over $1.4 million in community grants alone. These grants were given to the city and county governments of the potential sites (Kerr and Seidler, 1989). The U.S. House of Representatives appropriated

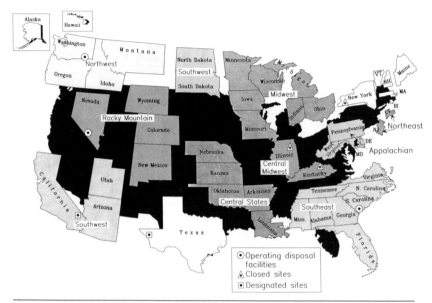

FIGURE 5-1. U.S. map showing compact alignments, the operating and closed LLRW disposal facilities, and the designated proposed sites in California and Texas. Unaligned states are shown in white. The existing sites are shown by circles; closed sites by triangles; and proposed sites by squares. The Northeast and Southwest compacts are not contiguous and will require travel outside their respective regions in order to transport LLRW from generators to the disposal facilities.

$5 million for the DOE to aid states and compacts in meeting the deadlines of the LLRWPAA (Helminski, 1989d). A study by the Electric Power Research Institute (EPRI) estimated that the cost of developing sites under the compact system will be approximately $5.4 billion (Helminski, 1989b).

Siting Process

The siting processes for the states with newly designated sites were similar. First, a state agency or contractors earmarked areas that were large enough to host the site. Then these areas were evaluated according to Nuclear Regulatory Commission (NRC) and state criteria, examining such factors as the geological, hydrological, and geophysical characteristics of the areas, the surrounding population density, and possible transportation routes. The suitability requirements for disposal sites according to 10 CFR Part 61.50 are listed below.

1. The site must be such that it can be characterized, modeled, analyzed, and monitored.
2. The location should not fall within areas that will probably be affected by future population growth and development.
3. The site should not contain exploitable natural resources, such as ore or petroleum reserves.
4. The site should be well drained and free of frequent flooding or ponding. The site specifically may not be placed in a 100-year floodplain, coastal high-hazard area, or a wetland.
5. The site should have minimum upstream drainage so that the possibility of erosion or inundation from runoff is minimized.
6. The site may not be within a zone where groundwater intrusion is possible.
7. The site should not be in areas of high seismic, volcanic, erosion, or mass wasting activity.
8. The site should not be located near other facilities or activities that may adversely affect the site or that have emissions that make monitoring the environmental impacts of the site difficult.

The evaluations included the review of environmental impact statements and responses at public hearings. During the evaluation period, prospective counties passed referenda banning disposal sites. What began as a choice between ten or more sites ended up being a choice between a handful of localities that had not passed such legislation. Generally the few choices left were in remote, sparsely populated, economically depressed regions. The state offered economic incentives to communities that were amenable to hosting a site. The NRC evaluated the site to ensure that it met the requirements for a license, and the site was designated. Generally, the site-selection process takes about 2 years, the preparation of the environmental impact statement another 2 years, and at least another 2 years is needed to construct the facility (Shaw, 1988).

As of 1990, one state and four compacts have designated sites. Six states and five compacts still face the task of designating sites. Land use restrictions complicate the process and are of particular impact in areas of the country that are not geologically or climatologically sited. For those areas that are suitable, the public assertion of "not in my backyard" and organized antinuclear activism make the site selection decision difficult. In the search for a 60-acre site, Illinois initially considered 21 different counties. By the time the state siting commission had determined that 17 of them were acceptable, nine counties had passed referenda prohibiting the establishment of an LLRW facility. The remaining eight counties passed similar referenda before the study was completed. They are now reduced to two locations (personal communication, David Ed, Illinois Department of Nuclear Safety). Chem-Nuclear, Inc., which operates a successful site in South Carolina, has suffered two failures in attempts to designate sites in South Dakota and Colorado. Compacts may choose a state to host its site and accept the compact's LLRW against the state's wishes. The host state must meet all its own regulations while the guidelines for the site are being developed and enforced by the compact as a whole. Some of these guidelines, established by agencies consisting of appointed, not elected, officials, may be in conflict with the regulations or laws of the host state. Illinois withdrew from the Midwest Compact negotiations when it recognized the probability of being chosen to host the site, and instead aligned itself with Kentucky, which produces very little LLRW. New York, Maryland, and Pennsylvania withdrew from negotiations to form the Northeast Compact, each to avoid being designated as the host state (NYSDEC, 1984). Both Nebraska and Kansas tried to withdraw from the Central States Compact negotiations for the same reason. North Carolina, after being designated by the Southeast Compact to host its future site, also attempted to withdraw (Schmidt, 1986).

Because LLRW generators have always had to ship their waste to a disposal facility, problems and risks associated with the transportation of LLRW are not new. However, the formation of compacts has introduced a new twist. For example, the Northeast Compact is planning to establish two disposal sites—one in Connecticut and the other in New Jersey—which will require transportation of wastes through New York State. Although supported by DOT legislation, this route will undoubtedly be met with protest from both New York State and New York City. In fact, Congressman Bill Greene from New York has already introduced the Radioactive Materials Transportation Protection Act of 1989 (HR2387), which, if passed, will restrict the transport of radioactive materials through high-density metropolitan areas. As the compacts designate sites, other transportation difficulties will arise, ranging from the necessity of building highway access roads to the establishment of emergency plans and tortuous truck routes to avoid densely populated areas. If compacts and states fail to designate sites, other problems will occur. For example, although treatment facilities may still accept LLRW from states that have not complied with the Amendments Act, the waste will have to be segregated and shipped back to the generator after compaction or incineration. This will double the distance that the LLRW is transported, increase costs and create logistical problems.

Assuming that all the compacts and nonaligned states meet the 1993 deadline to establish disposal sites, the costs for disposal are expected to increase dramatically. Volume-reduction methods and the use of alternative techniques already have led to the production of less LLRW as shown in Figure 1-13. If this trend continues, it will force disposal facilities to raise their charges. In order to avoid this, some compacts are planning to establish a minimum disposal cost based upon the current volume produced (personal communication, David Ed, Illinois Department of Nuclear Safety). In this way, the abundance of sites will discourage the implementation of additional volume reduction methods. The price for disposal in 1989 is approximately $40/ft^3 ($1400/m^3); development costs and the expensive methods being considered threaten to double or even quadruple this cost. In the end, considering the high construction and operating costs, many of the new sites may not be economically viable.

During the next 5 years, the states and compacts will be responsible for financing the agencies and commissions that will develop their sites. Several plan to use the surcharge rebate monies that they will receive as a result of compliance with the 1988 deadline; others plan to levy annual fees against member states and generators. Generally, the compacts are not planning to finance the development costs and are instead requiring the host state to foot the bill. In the nonaligned states, the monies will be raised from license fees, asssessments on nuclear reactors, and general appropriations. Most states' agencies are acting under legislation that guarantees recovery of expenses for facility development, monitoring, and regulation. Regardless of the financing plan, ultimately it is the generators, and therefore the consumers, who will pay for the sites.

Deadlines

The January 1, 1988, deadline of the Amendments Act required all compacts to have named their host states, and all unaligned states to have shown themselves ready to develop a site. The four states that failed to comply—Maine, New Hampshire, North Dakota, and Vermont—were subject to a doubling of the surcharge through July 1988, and a quadrupling after that date. The compliance states were eligible for rebates, representing a percentage of the surcharges they have paid to date (Blake, 1988). Vermont and New Hampshire failed to correct their situations by January 1, 1989, and have consequently been denied access to the three operating commercial sites. Generators in Vermont, who produced a total of 205 m^3 in 1988, 95% of which came from the Vermont Yankee nuclear power plant, and those in New Hampshire, who produced only 5 m^3, must store their LLRW on-site until they develop other options.

Considering that the Amendments Act is a result of the states' failure to meet the deadlines set forth in the Policy Act, the following dates for upcoming deadlines may not be absolute, but the progression will most likely remain the same (Table 5-1). Based on their progress to date, it is unlikely that all the states will meet the deadlines. By

TABLE 5-1. Deadlines Defined by the Low-Level Radioactive Waste Policy Amendments Act of 1985

Date	Legislated Action(s)
January 1, 1986	Each state to have joined compact or to have enacted legislation indicating intention to develop its own site; surcharge not to exceed $10 per cubic foot.
July 1, 1986	Generators in states that did not meet the January 1, 1986 deadline are subject to doubled surcharges until December 31, 1986.
January 1, 1987	Generators in states that did not meet January 1, 1986 deadline may be denied access to operating disposal sites.
January 1, 1988	Compacts to have named host states; unaligned states to have developed siting plan and schedule and to have delegated authority for development; surcharge not to exceed $20 per cubic foot; noncompliance states subject to doubled surcharges.
July 1, 1988	Noncompliance states subject to quadrupled surcharges.
January 1, 1989	Generators in states and compacts that did not meet the January 1, 1988 deadline to be denied access to operating disposal sites.
January 1, 1990	Compacts and unaligned states to file a complete operating license application; letter from governor stating that the unaligned state will have provisions for LLRW disposal in place by December 31, 1992, may be submitted in lieu of application; surcharge not to exceed $40 per cubic foot; failure to comply may result in denial of access to operating disposal sites.
January 1, 1992	All compacts and unaligned states to file operating license applications; letter from governor no longer sufficient for compliance status.
January 1, 1993	Sited compacts to be empowered to restrict import of non-compact LLRW.
January 1, 1996	Surcharge rebates cease.

January 1, 1990, all unsited states must have either filed a complete application for an operating license for a designated disposal site (either independently or through a compact), or the governor must assert in writing to the NRC that the state will be capable of managing its LLRW after December 31, 1992. Compliance is recognized immediately after a license application is filed. A copy of the application must be sent to all sited states. If compliance is vouchered through submission of a letter from the governor, the letter must include a projection of how much LLRW will be produced by the state after 1992, listed by class and generator type. The letter must also address which methods will be employed for storage and management, which agencies will have authority over the scheduling and development of the facility, and provide assurance that the plans will comply with the NRC and other applicable regulations (Helminski, 1989a). Compliance states will review the letter and rule as to whether it is sufficient to meet the deadline requirements. Noncompliance with this deadline can lead to continued denial of access to existing disposal sites. By January 1, 1992, a license application must be filed. The generators in states without sites may be subject to additional surcharges assessed by the state receiving the waste—$20/ft^3 ($707/m^3) in 1988–89, increasing to $40/ft^3 ($1413/m^3) in 1990–92. This means that after 1990, these generators will be paying, on average, as much in surcharges as they are for actual disposal. After January 1, 1993, the sited compacts will be empowered to refuse LLRW from other non-compact-member states and many have already stated that they will do so. Surcharge rebate payments to states and LLRW generators who still have not developed a site but are seeking to develop one will cease after January 1, 1996. After 1996, no schedule of penalties or rebates exists. The Congress will reevaluate the compacts every 5 years. For those states that decide simply to defer any decision, the consequences may be the forfeiture of surcharge rebates, increased surcharge penalties for the generators, and exclusion from some of the sites. If the volume of waste continues to decrease, existing sites may continue to accept wastes at higher and higher prices. This increase in cost, paid by the generator, will either be passed on to the public or will result in suspension of services.

Despite how straightforward the deadlines may appear when listed as above, the compacts and especially the unaligned states are having problems in meeting and implementing the requirements. The Amendments Act is biased against single states developing their own sites since, by not joining a compact, unaligned states are in violation of the act. Even if they meet all the deadlines, the act may not grant them the privilege of excluding imported waste. Individual states may contest the ability of any state or compact to exclude imported waste or restrict export of waste, despite what the Amendments Act states, based on the supremacy and commerce clauses. New York State is attempting to circumvent this problem by developing a state-owned and operated facility that may be exempt from these clauses. According to the siting plans submitted to the DOE, only three states (Texas, California, and Illinois) will meet the 1990 deadline and only two more will be able to meet the 1993 deadline (North Carolina and Nebraska). It appears that New York, Pennsylvania, Michigan, Connecticut, New Jersey, Maine, and Massachusetts will not meet the 1993 deadline (Helminski, 1988c, 1989d). The Maine Low-Level Radioactive Waste Authority has initiated development of a storage facility, operational in 1993, to manage the waste until a disposal facility can be developed. The single nuclear power plant in Maine (Maine Yankee), which produces 95% of the state's LLRW, is financing the development (Scott et al., 1989).

Compacts

The current status of the compacts and unaligned states is shown in Table 5-2. The three operating commercial disposal facilities will serve as the first sites for their respective compacts. Not surprisingly, these three compacts were the first to form after the Policy Act was passed. The presence of existing sites has eased the burden of meeting the deadlines for these three compacts. The Northwest Compact has yet to designate where its next facility will be located once the Richland site closes. Colorado will become the new host state for the Rocky Mountain Compact at the end of 1992, and North Carolina will host, against its wishes, the next site for the Southeast Compact beginning in early 1993. North Carolina has announced potential sites in Richmond, Rowan, Union, and Wake/Chatham Counties.

Compacts requiring the development of new sites understandably lag behind compacts with established sites in meeting many of the deadlines. The host state for the Appalachian Compact is Pennsylvania. Pennsylvania is postponing the search for a site until a disposal method is Chosen and the design is developed. Illinois, host state for the Central Midwest Compact, is currently performing a study of two potential sites, one in Clark County and one in Wayne County. The Central States Compact has designated Nebraska as host state. In a statewide referendum in November 1988, the citizens of Nebraska

TABLE 5-2a. Status and Plans of the Compacts

Compact	Member States	Host State	Site Selection
Northeast	Connecticut New Jersey	Connecticut New Jersey	Not chosen
Appalachian ratified 1985	Delaware Maryland[1] Pennsylvania[1] West Virginia	Pennsylvania	Not chosen
Southeast ratified 1985	Alabama[1] Florida[1] Georgia[1] Mississippi[1] N[1] & S Carolina[1] Tennessee[1] Virginia	South Carolina until 1993 North Carolina[2] after 1993	Barnwell, SC Four possible locations
Central States ratified 1985	Arkansas[1] Kansas[1] Louisiana[1] Nebraska[1] Oklahoma	Nebraska[2]	Nemaha County, or Nuckolls County, or Boyd County
Midwest ratified 1985	Indiana Michigan Minnesota Missouri Iowa, Ohio Wisconsin	Michigan	To be chosen in 1990
Central Midwest	Illinois Kentucky[1]	Illinois[2]	Clark County or Wayne County
Rocky Mt. ratified 1985	Colorado[1] Nevada[1] New Mexico[1] Wyoming	Nevada until 1992 Colorado[2] after 1992	Beatty, NV Uravan, CO
Southwest ratified 1988	Arizona[1] California[1] N[1] & S Dakota	California[2] Arizona after 30 years	Needles, Ward Valley
Northwest ratified 1985	Alaska Hawaii Idaho[1] Montana Oregon[1] Utah[1] Washington[1]	Washington until 1992 Subsequent host not chosen	Richland, WA

[1] Agreement State
[2] Possess Agreement State Authority to regulate commercial LLRW

Proposed Method	Regulatory Agencies	Developer
No SLB	CT Hazardous Waste Mgt. Service NJ LLRW Siting Board Northeast Compact Commission	Not chosen
AGV	PA Bureau for Rad. Protection PA Dept. of Env. Resources	US Ecology or Chem-Nuclear
SLB	SE Compact Commission NC LLRW Authority	Chem-Nuclear
Not chosen	Central States Compact Commission	US Ecology or Bechtel National
No SLB	Michigan LLRW Authority Midwest Compact Commission	Not chosen
EMCB	Illinois Dept. of Nuclear Safety Central Midwest Compact Commission	Westinghouse-design Chem-Nuclear-development/ operation
SLB Mined cavity	Rocky Mt. LLRW Board Colorado Dept. of Health	US Ecology
SLB	CA Dept. of Health Services	US Ecology
SLB	WA Dept. of Ecology	US Ecology

TABLE 5-2b. Status and Plans of the Unaligned States

Compact	Member States	Host State	Site Selection
Unaligned	Maine	Seeking compact membership	Not chosen
	Massachusetts	Seeking compact membership	Not chosen
	New Hampshire[1]	Seeking compact membership	Not chosen
	New York[1]	New York[2]	Five possible locations West Valley banned
	Rhode Island[1]	Seeking compact membership	Not chosen
	Texas[1]	Texas[2]	Hudspeth County
	Vermont	Seeking compact membership	None
	Washington, DC	None	Agreement with Rocky Mt.

[1] Agreement State
[2] Possess Agreement State Authority to regulate commercial LLRW

voted (64 to 36%) to stay in the Central States Compact. Since then the compact commission has chosen three potential sites in Nemaha, Nuckolls, and Boyd Counties. The Midwest Compact has designated Michigan as its host state, and Michigan has established a siting agency headed by a single individual who has sole authority over site and design selection. Michigan is searching for a 1200-acre site in either St. Clair, Ontonagon, or Lenawee County. Selection is to be made in early 1990. Connecticut and New Jersey will both host sites for the Northeast Compact. California will be the first host for the Southwest Compact with a facility in the Mojave Desert near Needles in Ward Valley.

Unaligned States

Those that had not entered into compacts as of January 1, 1988, were seven states, the District of Columbia, and all other U.S. territories. With regard to the act, the territories are considered to have the same status as states. Of the seven states, Texas, New York, and Massachusetts have decided to develop sites themselves. Texas, which is being pursued by Vermont, Rhode Island, Maine, and Puerto Rico as a compact partner, has designated a site. New York is still in the process of finding a site but has identified five locations in Cortland and Allegany counties: two in the town of Taylor and one each in

Proposed Method	Regulatory Agencies	Developer
No SLB	Maine LLRW Authority	Not chosen
AGV or BGV	LLRW Management Board	Not chosen
Not chosen		
AGV, BGV, MC No SLB	NYS LLRW Siting Commission NYSERDA, NYS DEC	Not chosen
Not chosen		
BGV None	Texas LLRW Disposal Authority VT Advisory Commission on LLRW	Not chosen None
None		None

Cancadea, Allen, and West Almond. Massachusetts had passed a referendum banning a disposal site within its borders. This referendum was found unconstitutional by the state supreme court, and Massachusetts is now in the siting process (Colglazier and English, 1988). Although Maine has initiated a process for site selection, the other sited states considered Maine to be out of compliance with the January 1, 1988, deadline. Rhode Island has established a 2-year contract with the Rocky Mountain Compact that is valid through 1989, under which their generators pay all the usual fees and surcharge plus a $20/ft^3 ($706/m^3) fee to the Rocky Mountain Compact. Vermont has stated that it will *not* comply with the act but does have a legislative commission looking into the LLRW disposal problem. Although New Hampshire wants to petition for access to join an existing compact, the state legislature did not accomplish this before the January 1, 1988, deadline. Washington, D.C. has established an access agreement with the Rocky Mountain Compact, the U.S. Virgin Islands are not in compliance, and Guam is considering petitioning for access to an existing compact (Blake, 1988).

Even though the responsibility for LLRW disposal may legally rest with the states, the problem of LLRW disposal is a national one. Generators are located in all parts of the country and, for the most part, the benefits are shared directly or indirectly by the entire population. In the present political climate, the major generators (Figure 1-14) have the most incentive to take action. To some extent, that pattern has already been set. Those states generating the most LLRW assume they will be nominated to host sites, and, in response, they elect either to remain unaligned or align themselves with low-volume-producing states. The Southeast and Northwest Compacts together generate about half of the nation's LLRW while the Rocky Mountain Compact generates less than 0.2% (Helminski, 1988a). Interestingly, the Rocky Mountain Compact is negotiating with Vermont, New Hampshire, Maine, Rhode Island, and the District of Columbia for their LLRW in exchange for a $50/ft^3 surcharge, beginning January 1990 until 1992 (Helminski, 1989e). States with insufficient waste volumes of their own will continue to seek wastes from other states. It is also likely that compacts will begin competing with each other to acquire the additional volumes of LLRW needed for economically viable operation of their facilities. Conversely, generators in states like New York, where there is no legal requirement to use an in-state facility, may begin to ship their wastes to the lowest bidder. A single site for the Appalachian and Northeast Compacts and the northeastern states (which includes six of the unaligned states) would receive about one-third of the nation's waste. If only one site were established for the Midwest and Central states area, it would receive most of the remaining waste. The establishment of four regional sites (Northwest, Midwest, Southeast, and Northeast) would shorten transportation and provide locations for centralized waste processing. However, the addition of just two new sites, even without further volume reduction, would cut the volume received at Barnwell and Richland in half, thereby extending the capacity but reducing the income of these two sites.

Proposed Technologies

Aside from the ocean dumping of LLRW employed during the first two decades of its nuclear age, the United States has extensive experience only with shallow land burial (SLB) disposal technology. Yet two other technologies, below-ground vaults and aboveground vaults, are the principal methods being considered for new facilities. SLB has been prohibited by the Appalachian, Midwest, Northeast, and the Central Midwest Compacts as well as by Maine, Massachusetts, and New York (Assistant Secretary for Nuclear Energy, 1987; NYSDEC, 1987a). Despite the fact that NRC studies and the commercial sites' histories have shown that SLB is a safe option, in deference to public perception, states tend to choose more highly engineered (albeit not safer) disposal technologies that are more expensive and result in higher exposures to workers.

Following a failure to site a facility in McMullen County (Colglazier and English, 1988), Texas is developing a 200- to 250-acre site near Fort Hancock in Hudspeth County in the west Texas desert, with a tentative operational date in 1991. The state is currently considering a facility that uses below-ground concrete vaults for highly irradiating Class A, B, and C wastes, and modular concrete canisters for low irradiating Class A, B, C, and mixed wastes. Texas estimates that the facility could accept 100,000 ft^3 (2,832 m^3) of LLRW and 1000 ft^3 (28.32 m^3) of mixed waste per year (Baird et al., 1989). Predevelopment and construction costs are estimated at $16 to 20 million, for a total lifetime cost of $200 to 300 million, resulting in disposal charges of $80 to 100/ft^3 ($2825 to 3531/m^3). El Paso County has spent over $1.6 million in tax monies fighting the location of the site (Helminski, 1989f).

California has designated 50 acres in the Mojave Desert near Needles, Ward Valley, for site development of a standard SLB facility. The arid conditions of the Mojave Desert have led to its selection for other invasive uses in the past—the erosion damage from military maneuvers during World War II is still evident. The fact that this area was not "pristine" resulted in a greater acceptance of the facility. The state has estimated that it will spend $6 million on predevelopment efforts. US Ecology will operate the California site. In addition to complying with all the required steps associated with the development, the company must also protect the desert tortoise, an endangered species that lives in the region (Wilson, 1988).

Illinois, the host for the Central Midwest Compact, is planning a 1000-acre site, the outer 700 of which will be a buffer zone. The method for disposal will be a multiple-engineered earth-mounded concrete bunker. Class A, B, C, and mixed wastes will all be placed in the same types of concrete modules. The modules will then be filled with cement grout and placed within reinforced concrete vaults. An impermeable plastic cover and an earthen cap will cover the vault (Holland et al., 1989).

The New York State LLRW Siting Commission is searching for an area of 400 to 1500 acres and will consider three possible disposal methods: below-ground vaults, above-ground vaults, and underground mined repository disposal (Orazio et al., 1988). The NRC has stated that they will provide minimal effort for studies on the last technology. The siting commission estimates that disposal costs at the new facility

will be \$80 to $120/ft^3$ (\$2825 to $4237/m^3$). After the site has been
selected, the New York State Energy Research and Development
Authority will assume responsibility for its development. The state
already owns an established facility and over 3000 acres at West Valley
in which multiple sites could be designated. The entire West Valley
facility has undergone extensive environmental characterization and
would meet the siting criteria for the majority of the disposal methods
listed above. Currently the DOE is disposing of LLRW here at what
had been the NRC-licensed fuel-reprocessing site at West Valley (Helminski, 1986). The DOE's lease at West Valley expires in 1992, after
which title to the waste may revert to New York State. Ironically, in
a political compromise, New York State has specifically ruled out
siting its new facility within the existing West Valley complex (New
York Senate Assembly, 1986; NYSDEC, 1987b). As early as September of 1988, New York announced that it would not meet the 1993
deadline and would therefore need to develop an interim storage plan.
West Valley was excluded from consideration for location of the interim storage facility as well. Just before the 1990 deadline, New York
announced that its interim storage plan will require generators to store
their own waste for an indefinite period. Although regulations governing interim storage may make this option prohibitively expensive
for all generators, this plan (and the problems associated with interim
storage) will most certainly create hardships for generators in densely
populated urban areas, where space is at a premium. Any plan for
interim storage must be reviewed and approved by the compliance
states.

The Appalachian Compact has decided to design its facility before
searching for a site. Placing high priority on retrievability, they are
developing an above-ground vault, an overengineered facility referred
to as a "Taj Mahal." After the design is complete, the Pennsylvania
Department of Environmental Resources will begin the search for a
site. Massachusetts is also planning to employ a concrete vault design
but has not decided whether it will be above- or below-ground. Nebraska, the host for the Central States Compact, has ruled that the
facility must be built above the existing grade level and must meet a
"zero-release" objective.

The other states planning to build sites are still in the selection
process and have not yet chosen disposal technologies. Only four of
the 11 siting plans submitted to comply with the 1988 deadline address
mixed-waste disposal (Carlin, 1988) and only four states—Tennessee,
Washington, Colorado, and South Carolina—currently have permits
to accept mixed wastes. The deadline to obtain a permit from the

EPA was September 1989. The compliance states decided that siting plans must address all LLRW, including mixed waste (Helminski, 1988b). Any state that might want to include incineration in their disposal plan must obtain permits from the EPA, NRC, and local authorities. Few places in the United States incinerate LLRW that is above *de minimis* levels because obtaining the permit is very difficult and because, to date, it has been less expensive to ship the waste for burial. Currently, no commercial power utilities incinerate LLRW (Loysen, 1988); only SEG and a dozen or so institutions incinerate LLRW (Cooley et al., 1981), doing so under special permit from the NRC.

Contractors and Site Developers

Who is siting, designing, and building the new facilities for the states and compacts? Only a handful of companies have experience with designing, developing, and operating commercial LLRW disposal facilities, and these companies exact high fees for fully repeating this cycle in each compact or compliance state. Dozens of consultants are involved in assisting in the siting process and developing facility design proposals. Although most of this money is derived from levied surcharges, some money comes from state treasuries.

Chem-Nuclear Systems, Inc., has extensive experience in running the Barnwell site, as well as in design and development efforts for sites in Colorado, New Mexico, Texas, and South Dakota. The North Carolina LLRW Authority and the Pennsylvania Department of Environmental Resources have both chosen Chem-Nuclear to build and operate their facility.

US Ecology, which operates the existing facilities in Richland, Washington, and Beatty, Nevada, together with Bechtel National, Inc., is designing and developing the new site in Nebraska. California has also chosen US Ecology to develop and operate its site in Needles.

Westinghouse is designing and siting a facility for Illinois, but will not actually build it. Illinois is negotiating a contract with Chem-Nuclear for the development and operation of the facility for the Central Midwest Compact. The Appalachian Compact is considering bids from both Chem-Nuclear and US Ecology for the design and development of the site in Pennsylvania.

Ebasco Services, Inc., is designing a prototype earth-mounded concrete bunker facility for the DOE. This prototype design will be submitted to the NRC to aid them in establishing the acceptance criteria for the license applications that will soon be submitted from the compacts (Eng et al., 1989). The NRC has very limited experience in reviewing LLRW disposal license applications and has stressed the need for cooperation between licensees and the regulators. In its most streamlined form, review of license applications will still require experts from 20 fields and take at least 15 months to complete. Even timely submitted applications may fail to receive approval before the 1992 deadline (Greeves, 1989).

Performance Objectives

The NRC defines the performance objectives for disposal sites in 10 CFR Part 61. In addition to meeting all NRC requirements, states and compacts must carefully weigh the cost of site development, operation, and maintenance against the long-lived liability of LLRW. Because insurance for the developers is not required by law, developers may proceed without this added expense. Even if they desire coverage, private insurance companies may choose not to assume the risks of the venture. Although generators remain liable for their waste even after disposal, the precedent set at Maxey Flats and Sheffield indicates that state taxpayers will ultimately bear the cost of site maintenance. The NRC requires a 100-year institutional control period, after which the site should not need active maintenance. Some states have mandated longer control periods. This requirement is in direct conflict with the desire of some states to provide for retrieval of the waste. Although the developers need to minimize their cost so that generators will be able to afford disposal, they also have to appease a public that is reluctant to accept disposal sites with even the most stringent safeguards.

A factor in the public's unwillingness to accept a LLRW disposal site is the perception that the site is unnecessary. Considering the number of new sites mandated by the act, the public is correct in this assessment. With too many new sites, individual states and compacts will find that they have no incentive for volume-reduction measures and that, instead of excluding imported waste, they will be seeking it, just to pay for the operational costs of the disposal facility. A study by the Electric Power Research Institute (EPRI) has shown that consolidating the existing compacts into four "supercompacts" and reducing the number of new sites to four would result in a $2.8 to 3 billion savings for the country (Helminski, 1989b). In the report that accompanied the allocation of $5 million to the DOE for its LLRW program, the Congressional House Appropriations Committee encouraged increased cooperation between compacts in order to reduce the number of sites developed (Helminski, 1989c). In an effort to highlight the need to reduce the number of sites being planned, Michigan governor Blanchard halted the siting process in Michigan and threatened to pull out of the Midwest Compact in February 1989. Blanchard charged that the Amendments Act was flawed and called for congressional action to force consolidation of the existing compacts. With no support from congressional leaders or other governors, and faced with denial of access to the three operating sites, Blanchard had no choice but to reinitiate the siting process. According to Michigan senators John Engler and Vern Ehlers, "Congress appears to have little interest in revisiting the federal law." Consolidation of the compacts will require a few of the host states to accept more LLRW. Even with federal guidance and incentives, it is unlikely that any of the host states will take on that added burden. On scientific and economic grounds, the development of as many as 16 sites for LLRW disposal is unjustifiable.

References

Assistant Secretary for Nuclear Energy. 1987. Report to Congress in Response to Public Law 99-240. In *1986 Annual Report on Low-Level Radioactive Waste Management Progress*. Washington, D.C.: DOE.

Baird, R. D., N. Chau, G. Merrell, V. Rogers, L. Oyen, and R. A. Alvarado. 1989. Preliminary design of LLRW disposal facility for the state of Texas. *Waste Management '89* 2:71–77.

Blake, M. E. 1988. LLW siting: Is there order within the chaos? *Nuclear News* March, 48–52.

Carlin, E. M. 1988. Mixed waste management in Washington and the Northwest Compact Region. *Waste Management '88* 1:935–938.

Colglazier, E. W., and M. R. English. 1988. Low-level radioactive wastes: Can new disposal sites be found? In *Low-Level Radioactive Waste Regulation: Science, Politics, and Fear,* ed. M. E. Burns, Chelsea, Mich.: Lewis Publishers, pp. 215–238.

Cooley, L. R., M. R. McCampbell, and J. D. Thompson. 1981. *Current Practice of Incineration of Low-Level Institutional Radioactive Waste.* EG&G-2076. Springfield, Va.: NTIS.

Eng, R., P. Liu, I. Tsang, and W. Chang. 1989. Earth-mounded concrete bunker disposal system. *Waste Management '89* 2:101–106.

Greeves, J. 1989. Licensing a LLRW facility. Paper read at *Waste Management '89*, 26 February to 2 March 1989, Tucson, Arizona.

Helminski, E. L. 1986. Westinghouse-West Valley set to start up to nation's newest LLRW disposal site. *Radioactive Exchange* 5(15):1–2.

Helminski, E. L. 1988a. LLRW volume disposal update, *Radioactive Exchange* 7(8):7.

Helminski, E. L. 1988b. Unsited states' '90 LLRW burial site applications must cover all waste. *Radioactive Exchange* 7(18):1–2.

Helminski, E. L. 1988c. DOE LLRW report estimates disposal costs as high as $113 per cu. ft. *Radioactive Exchange* 7(18):5.

Helminski, E. L. 1989a. Sited states issue LLRWPAA 1990 milestone certification criteria. *Radioactive Exchange* 8(2):3–4.

Helminski, E. L. 1989b. EPRI study sees $2.8 billion savings if compacts are consolidated to four. *Radioactive Exchange* 8(6):1.

Helminski, E. L. 1989c. House gives LLRW $5 million; wants a LLRW group to promote less sites. *Radioactive Exchange* 8(11):1–2.

Helminski, E. L. 1989d. New York Interagency Task Force. *Radioactive Exchange* 8(16):6.

Helminski, E. L. 1989e. Rocky Mtn. charges VT, NH, ME, RI, DC $50 per cu. ft. to take LLRW. *Radioactive Exchange* 8(16):4.

Helminski, E. L. 1989f. Texas LLRW Authority selects disposal site in Hudspeth County. *Radioactive Exchange* 8(20):1–2.

Holland, J., D. Hoffman, and D. Meess. 1989. Engineering a low-level radioactive waste disposal facility: The Illinois design. *Waste Management '89* 2:79–81.

Kerr, T. A., and P. E. Seidler. 1989. Achieving local support for a low-level radioactive waste disposal facility in Illinois. *Waste Management '89* 2:43–47.

Loysen, P. 1988. NRC's role in LLW incineration. In *Proceedings of the Annual International Conference on Incineration of Hazardous, Radioactive & Mixed Wastes,* San Francisco, California, p. 4.

New York Low-Level Waste Management Act, New York Senate Assembly S. 9616, A. 11729, 1986.

New York State Department of Environmental Conservation (NYSDEC). 1984. *Low-Level Radioactive Waste Management Study, Volume One; Executive Summary.* New York: NYSDEC.

New York State Department of Environmental Conservation (NYSDEC). 1987a. *Recommendations for State Assistance to Localities Affected by the Siting of a Low-Level Radioactive Waste Management Facility.* New York: NYSDEC.

New York State Department of Environmental Conservation (NYSDEC). 1987b. *6 NYCRR Part 382: Regulations for Low-Level Radioactive Waste Disposal Facilities.* New York: NYSDEC.

Orazio, A. F., W. F. Schwarz, and A. X. Feeney. 1988. New York's response to the national LLRW disposal legislation. *Waste Management '88* 1:55–59.

Schmidt, W. E. 1986. North Carolina is selected for a nuclear waste dump. *The New York Times,* September 12, 1986.

Scott, M., S. N Thompson, W. A. Anderson, and J. S. Williams. Status of Maine's low-level radioactive waste program. *Waste Management '89* 2:40–42.

Shaw, R. 1988. Low level radioactive wastes in the nuclear power industry. In *Low-Level Radioactive Waste Regulation: Science, Politics, and Fear,* ed. M. E. Burns, Chelsea, Mich.: Lewis Publishers, pp. 119–140.

Wilson, D. S. 1988. And the winner gets . . . a radioactive dump. *The New York Times,* March 20, 1988.

6 RISK AND RADIATION

We tend to think in absolute terms when our safety is concerned. Something is either safe or unsafe. To many, "risk" is the range of probability between zero and certainty. If we are not certain that something is safe or unsafe, we say it is "risky," and when we use the word "risk," we usually imply that there is some uncertainty. Understanding the concept of risk, the regulatory approach to radiation, the relationship between the amount of radioactivity in low-level radioactive waste (LLRW) and what is a natural part of our environment, the risks associated with low doses of radiation, and the management of risks associated with LLRW is central to solving the LLRW problem.

Concept of Risk

A review of the federal regulations governing risk shows that we have been inconsistent in accepting a given level of risk (Ames et al., 1987; Berg and Maillie, 1981; Byrd and Lave, 1987; Covello, 1987; Lave, 1987; Lowrance, 1983; NCRP, 1980; NCRP, 1981; Okrent, 1987; Perera, 1985; Russell and Gruber, 1987; Slovic, 1987; Wilson and Crouch, 1987). Also, we have not always considered the impact of losing the benefits of the risk-associated activity. With considerable public support, the Congress has expressed this inconsistency by accepting industry standards of 1 premature death in 10,000 per year (NSC, 1987) (higher in specific situations regulated by the OSHA, FDA, and TSCA), yet requiring a reduction in risk, no matter how small, in the Food Additive Amendments Act. For example, the FDA banned saccharin from food products and then reversed its decision under congressional response to public pressure for an artificial sweetener (*Federal Reporter,* 1987). Following a Supreme Court decision in 1980 concerning a standard regulating occupational exposure to benzene (Supreme Court, 1980), federal regulatory agencies now must demonstrate a significant risk before they can regulate the exposure. Subsequently, the EPA (54 *Federal Register* 6392 February 10, 1989; 40 CFR Part 141), FDA (*Federal Reporter,* 1987), and NRC (10 CFR Part 2) have determined for policy purposes that 1 premature death in 1,000,000 for the public is too low to be of regulatory concern.

Usually, demonstration of risk precedes its regulation. However, under a specter of nuclear holocaust, this process has been reversed for low-level radiation, and the beneficial uses of radioactivity have not been fully appreciated. We invariably think that any radioactivity poses a greater risk to health than smoking or driving, despite evidence to the contrary. This perception has resulted in complex, and often overly stringent, regulations. Clearly, psychological, social, economic, and political factors play substantial roles in determining how different risks will be managed.

Regulatory Concepts

More is known about the health effects from exposure to radiation than any other physical or chemical hazard. A very large body of data has been collected and risk assessments have been performed by various national and international councils. It is reassuring that their recommendations regarding low-level radiation protection have been in general agreement. Principally, these groups are the National Academy of Sciences' Committee on the Biological Effects of Ionizing Radiations (BEIR), the United National Scientific Committee on the Effects of Atomic Radiation (UNSCEAR), the National Council on Radiation Protection and Measurements (NCRP), the International Commission on Radiological Protection (ICRP), the International Atomic Energy Agency (IAEA), and the Nuclear Regulatory Commission (NRC).

Several conceptual terms, that is, MPC, MPD, ALARA, BRC, and *de minimis,* are used to guide the management of low-level radiation (Figure 6-1) (Alexander, 1988; Auxier and Dickson, 1983; Byrd and Lave, 1987; Comar, 1979; Eisenbud, 1987; IAEA, 1987; Kocher, 1986; Mumpower, 1986; NCRP, 1987e; Rogers and Murphy, 1987; Schiager et al., 1986; 10 CFR Parts 2 and 20). The maximum permissible dose (MPD) levels, assuming lifetime radiation exposure, are set conservatively below those at which health effects have been observed. Gen-

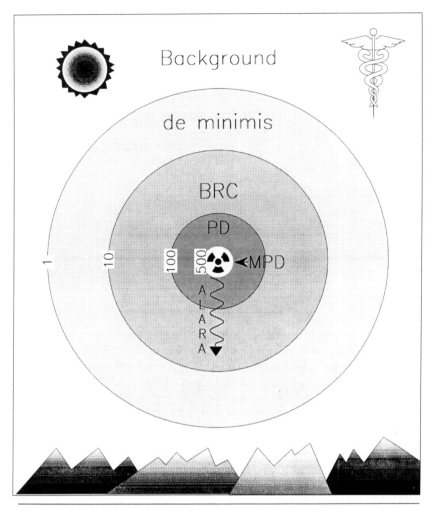

FIGURE 6-1. Common concepts for regulation of radiation. Although cosmic and geological (principally radon) background sources are the major sources of radiation exposure, medical diagnoses are the largest source of manmade exposure. The numbers represent doses expressed in mrem.

erally, standards for specific practices are below the maximum limit of acceptable dose from all combined sources. The NRC's regulatory concept of keeping dose "as low as reasonably achievable" (ALARA) (10 CFR Part 20.1) means that dose pathways must be both below the MPD and minimized to the extent that doses approach an insignificant value (see *de minimis* dose below), cost, and feasibility per-

TABLE 6-1. Comparison of Recommendations
for Deregulation

	de minimis Dose/yr	BRC Dose/yr	
	Individual	Individual	Collective
NRC	1 mrem	10 mrem	Low
EPA	....	4 mrem	Low
ICRP	1 mrem	....	....
IAEA	....	1 mrem	100 person-rem
NCRP	1 mrem	....	....
HPS	....	10 mrem	None

mitting. Conceptually related to the MPD is the maximum permissible concentration (MPC) of specific radionuclides that may be released to the environment as gaseous or liquid effluents (10 CFR Part 20 Appendix B).

A *de minimis* dose defines a range of exposure below which health physicists think no quantifiable risks exist. *De minimis* means the least and is an abbreviation for *de minimis non curat lex,* generally translated "the law does not pay attention to the trivial." Even though exposures in the range of 1 to 30 times the average natural background level (Figure I-1) have not provided any evidence of adverse health effects, that is, increased incidence of cancers, in practice, *de minimis* levels of radiation are generally set at only a fraction of the 364 mrem/yr (3.64 mSv/yr) contributed by the various natural sources of radiation. Because levels for all types and amounts of radionuclides potentiating a dose below 1 mrem/yr (0.01 mSv/yr) are generally considered trivial (Alexander, 1988; Auxier and Dickson, 1983; IAEA, 1987; Kocher, 1986; NCRP, 1987e; Schiager et al., 1986; 10 CFR Parts 2 and 20), they should be formally recognized as *de minimis*. It is on the order of 0.03% of the average annual radiation dose normally absorbed by members of the public. Unlike BRC levels (see below), *de minimis* levels are generally defined solely in terms of probable mortality for exposed individuals, exclusive of the size of the exposed population and the total number of expected mortalities.

ALARA levels should be evaluated as the basis for establishing levels for disposal or environmental release of radioactivity that are above *de minimis* levels but "below regulatory concern" (BRC). BRC can be thought of as the lower limit of ALARA. Unlike *de minimis*, BRC levels are dependent upon the waste stream, disposal technologies available, and the potential for exposure. BRC levels are designated by conscious decision at values below which the benefits to society outweigh the risks. In 1988, the NRC suggested a limit for the maximum individual dose of 10 mrem/yr (0.1 mSv/yr) from a single source and 100 mrem/yr (1 mSv/yr) for cumulative exposure to individuals from multiple BRC sources. At the same time the EPA defined its criteria for classification of waste as BRC, which included a maximum individual dose of 4 mrem/yr (0.04 mSv/yr) (Holcomb et al., 1989), the same as its *de minimis* level for drinking water (40 CFR Part 141). Also, both agencies specified that the collective dose (see below) to the population must be small. The IAEA has recently published criteria for determining BRC quantities of specific isotopes in LLRW for disposal in municipal landfills or by incineration (IAEA, 1987); the limiting factor is the potential exposure to disposal-site workers, not to members of the public. A comparison of the levels recommended by different agencies is given in Table 6-1. To be effective, BRC rules must be extended to all states and localities and be accepted by other regulatory agencies which have overlapping authority, for example, HHS, FDA, EPA, DOT, OSHA, MESA, and DOL.

To quantify the effects of very low exposures, regulators have often adopted the epidemiological concept of collective dose. For radiation, collective dose is expressed in person-rems (person-Sieverts) and is the sum of the individual doses in the population. Although important to the decision-making process, the concept of collective dose can also be misleading. Since exposure to radiation can come from different sources, it is useful to be able to compare the effects of different radiation sources on different populations as well as to evaluate the total exposure individuals might receive in order to assure that it will be below the MPD. The main use of the collective dose is to predict health effects by extrapolating linear response relationships from high-level exposures to estimate the number of cancers potentially developing in the population. Critics argue that collective doses are impossible to measure and are usually just theoretical calculations. Also, the person-rem (person-Sievert) does not take into account the dose rate, that is, whether a higher exposure was received at one time or several lower exposures at different times. A collective dose for 1000 persons exposed to 10 mrem (0.1 mSv) is indistinguishable from 10 persons exposed to 1000 mrem.

Fundamental to understanding the health consequences of exposure to radiation is the dose rate or dose intensity. Consider that one might receive 25,000 mrem (250 mSv) during a lifetime [364 mrem

(3.64 mSv), the average annual dose received from background radiation in the United States, each year for 70 years]. What if that
dose were received over a year, month, day, minute, or shorter time?
As one might expect, in most biological systems, given that the total
dose remains the same, acute irradiation at a higher dose rate usually
results in an increased biological effect when compared to a chronic
dose at a lower rate (Figure 6-2). Therefore the accumulated dose as
well as the rate at which it is received are of interest. Although it has
generally been assumed that at low doses and low dose rates radiation
effects do take place but are not observed, it is also possible that a
given amount, or threshold level, of radiation is required for effects
to occur. Failure to observe health effects might be due to the ability
of living organisms to repair damage to genetic material (DNA), a
process that has been going on throughout evolution. Yalow (1989)
has proposed a model for radiation threshold effects:

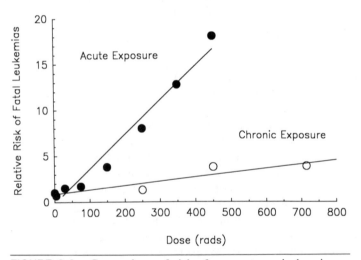

FIGURE 6-2. Comparison of risks from acute and chronic exposure to radiation. Comparing the risk from fatal leukemias in
the Japanese population after the bombings in World War II (Kato
and Schull, 1982) to the career-long occupational exposures of
radiologists (Matanoski et al., 1975) illustrates the difference between acute and chronic exposures to radiation (also see legend
to Figure 6-4). The doses received by radiologists were estimated
from the radiation standards in effect during their period of employment. Actual doses may have been lower or substantially
higher. This curve also attests to the efficiency of our progressively
restrictive radiation standards.

Since we are 75% water, low-LET [linear energy transfer] ionizing radiation is largely absorbed in the water resulting in the production of free radicals. Thus many of the potential biochemical changes initiated in the cell and, in particular, damage to cellular DNA are probably a consequence of the action of the products of water radiolysis. If molecules which scavenge radicals and which are normally present in tissue greatly exceed in concentration the free radicals generated at low dose rates, there may well be no initiating event, i.e., damage to DNA. The threshold could be the dose rate at which the radiation-induced free radicals exceed the scavengers.

She quotes from NCRP Report No. 43 (1975) that "The indications of a significant dose-rate influence on radiation effects would make completely inappropriate the current practice of summing of doses at all levels of dose and dose rate in the form of total person-rem (person-Sievert) for purposes of calculating risks to the populations on the basis of extrapolation of risk estimates derived from data at high doses and dose rates. . . ." Yalow (1989) has suggested that in place of the person-rem (person-Sievert) concept, we use "rem-person" and compare occupational and other manmade exposures to the radiation exposures acquired from natural sources.

The overwhelming majority of health and safety professionals consider doses in the range of 1 to 10 mrem/yr (0.01 to 0.1 mSv/yr) to be *de minimis* (Alexander, 1988; Eisenbud, 1987; IAEA, 1987; Kocher, 1986; NCRP, 1987e; Schiager et al., 1986). The potential dose to the public from most LLRW is in this range. Certain generators have been permitted to disregard radioactive releases below certain levels. For example, the DOE's Savannah River Plant uses a 10 μCi/g (370 kBq/g) cutoff limit for identifying some materials that can be disposed of as nonradioactive. The lack of similar *de minimis* levels for other LLRW generators means that, with few exceptions, as long as radiation can be detected, the waste materials must be disposed of as radioactive unless otherwise permitted. With the sensitive instrumentation available today, levels of detection are often well below natural background radiation, leading to the unnecessary disposal of waste as radioactive. This is costly and often does not permit proper treatment of any inherently hazardous, nonradioactive characteristics of the waste, such as chemical toxicity or infectivity. Moreover, failure to recognize that *de minimis* levels do in fact exist erroneously reaffirms our perception that all levels of radiation are hazardous. This perception applies to chemicals as well. We assume that if exposure to any amount of a chemical is toxic or carcinogenic, then every amount of that substance is dangerous. This perception is reflected in our regulation of hazardous chemicals. Once a chemical is regulated under RCRA as hazardous, it generally remains so, regardless of what treatment it has undergone. For example, the ash remaining after incineration of flammable chemicals must be specifically delisted or it will still be regulated as a hazardous waste even though no hazardous residues remain.

Much of the radioactivity in LLRW above *de minimis* levels (predominantly Class A) is still very minimal, and those knowledgeable in the field think it should be considered BRC (Alexander, 1988; Eisenbud, 1987, 1988; IAEA, 1987; NAS, 1980; NCRP, 1986; Rogers and Murphy, 1987). The EPA has estimated that deregulation of waste materials that produce doses below 4 mrem/yr (0.04 mSv/yr) could eliminate up to 35% of the LLRW now sent to LLRW disposal sites (Gruhlke et al., 1989). Reducing the volume of materials that must be disposed of as LLRW would decrease the need for multiple sites, extend operating life of existing sites, result in significant cost savings to generators, and permit resources to be reallocated to better serve our society. The difference between the views of the scientific community and the regulators results from the latter having to answer not only to the Congress but also to public opinion.

Environmental Radiation

In judging the risks associated with the disposal of LLRW, the potential impacts on the environment as well as on health must be considered. Disposal of LLRW, even though its contribution is small, could begin to increase the background radiation levels against which all other radiation exposures are compared. Sources of natural background radiation consist of cosmic radiation from outer space, such as the sun; terrestrial radiation, resulting from the presence of naturally occurring radionuclides in the soil and earth; and the deposition of naturally occurring radionuclides in the human body, principally ^{3}H, ^{14}C, ^{40}K, ^{209}Bi, ^{210}Pb, ^{210}Po, ^{222}Rn, ^{226}Ra, ^{232}Th, ^{235}U, and ^{238}U (Jacobs, 1968; NCRP, 1976, 1983, 1985a, 1987a, 1987b). A fairly constant inventory of these radionuclides is maintained in the environment. Contributors to the absorbed radiation dose as a result of their presence in food, drinking water, and air are ^{3}H, ^{7}Be, ^{14}C, ^{22}Na, and ^{40}K (NCRP, 1983, 1984, 1987b, 1987a). The average amounts of ^{14}C and ^{40}K present in the body are approximately 0.1 μCi (3.7 kBq) each (NCRP, 1985a; Shapiro, 1981). These radioisotopes are readily accessed by all living organisms depending on the molecules in which they are contained and how they are metabolized once incorporated. Except for ^3H, ^{14}C, ^{40}K, ^{131}I, ^{137}Cs, ^{210}Pb, and ^{210}Po, most of the ingestible radionuclides are poorly absorbed (Shapiro, 1981). Ingestion and inhalation (and for ^3H, absorption through the skin) of the long-lived radioisotopes, rather than external irradiation, are the principal sources of our exposure to the radioactivity from LLRW (BEIR, 1980, 1988; NCRP, 1987b; Shapiro, 1981). Of course, we are exposed to releases of radioactivity from common sources everyday. In New York City alone, the total ^{14}C and ^3H used in the many biomedical institutions per year is less than 10% of the naturally occurring ^{14}C in organic components of the collected garbage and only a few percent of the ^3H found in rainfall (Yalow, 1988a, b).

Tritium and ^{14}C, the principal radioactive components of nondecayable, non-fuel-cycle wastes, are ubiquitous, long-lived, and naturally occurring. They are continually created by interaction of cosmic rays with nitrogen in the atmosphere where they form radioactive carbon dioxide, water, and other compounds that cycle through the biosphere. Approximately 3625 MCi (1.34×10^{20} Bq) of ^3H (NCRP, 1979) and 270 MCi (9.99×10^{18} Bq) of ^{14}C (NCRP, 1985a) are in the global inventory. Atmospheric weapons testing is responsible for the major manmade releases of ^3H [3600 MCi (1.33×10^{20} Bq) before 1963 (NCRP, 1979)], ^{14}C [9.6 MCi (3.55×10^{17} Bq) (NCRP, 1985a)], ^{129}I [10 out of 40 Ci (148 GBq) (NCRP, 1983)] and ^{137}Cs [34 MCi (1.26×10^{18} Bq) before 1963 (NCRP, 1977)]. Nuclear power plants contribute approximately 20 thousand Curies (kCi) (740 TBq) (NCRP, 1979, 1983, 1985a; Tichler and Norden, 1986) of radioactive ^3H and ^{14}C per 1000 MW. Based on our survey of the major institutions in New York City and radiochemical manufacturers, only 1 kCi (37 TBq) of ^3H and 0.2 to 0.4 kCi (7.4 to 14.8 TBq) of ^{14}C per year is associated nationally with institutional LLRW. Industrial wastes generated during the production of these radiochemicals have approximately ten times as much activity.

Although some radionuclides in LLRW are not naturally occurring, their amounts are relatively small. For the naturally occurring ones, the world inventory is orders of magnitude greater than that from utility LLRW, which is in turn much greater than the inventory from institutional LLRW. If all of the radioactivity in the utility wastes were released to the atmosphere, the change in potential exposure would be very small, even without uniform distribution and dispersion. For institutional wastes, this change would be far below our levels of detection. Of course just because we cannot detect it does not mean we should release it. However, the consensus remains that such a trivial increase in background radiation would have no effect on our health or the environment.

Risk from Low Doses

The average annual dose per person from ionizing radiation in the United States is about 364 mrem (3.64 mSv) of equivalent whole-body exposure (Figure I-1) (BEIR, 1988; NCRP, 1987a, c, d, NLLRWMP, 1980). Of this, approximately 82% is from natural background radiation (55% from radon), 15% from medical sources, 3% from consumer products, and less than 1% from occupational, fallout, and fuel-cycle sources. The earlier fears that exposure to radiation would spawn genetic defects has vanished except for its popularity in science fiction. The consensus among epidemiologists is that cancer induction is the principal potential health hazard associated with exposure to low levels of radiation, regardless of source (Boice and Fraumeni, 1984; BEIR, 1980; Comptroller General of the United States, 1981). Unfortunately, we tend to think that the major causes of cancer are manmade chemicals and radiation. This is clearly not true, since cancer rates in the United States, with the exception of lung cancer (from smoking) and breast cancer (etiology unknown), and some occupationally related cancers, have continued to decline since the last century (Riggan et al., 1983; Silverberg and Lubera, 1986). Since radiation effects at low doses cannot be measured directly, this information is extrapolated from studies of people exposed to high levels of radioactivity, e.g, patients undergoing X-rays and radiation therapy, Hiroshima and Nagasaki survivors, and people facing occupational exposures, such as painters of radium watch dials, uranium miners, and radiologists (BEIR, 1980; Boice and Fraumeni, 1984; Comptroller General of the United States, 1981; Kato and Schull, 1982; Kumazawa et al., 1984; Matanoski et al., 1975; NCRP, 1985b; Tirmache, 1988). Based on cause-specific mortality among atomic bomb survivors, a constrained, linear quadratic relationship between dose and effect has been assumed (Figure 6-3) (BEIR, 1980; NCRP, 1982). We have chosen fatal leukemias because they are easy to diagnose, have a short latency, and were observed in a large cohort (approximately 30,000). The curve, showing a linear relationship above the 10-rad (0.1-Gy) range, has been plotted by linear regression. Note that the number of leukemias attributed to other causes (dose = 0 rads) were in the same range or greater than those in the cohort exposed to 1 to 20,000 mrad

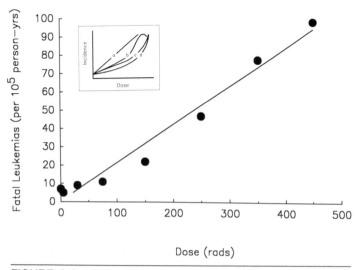

FIGURE 6-3. Relationship between the incidence of leukemia and dose among survivors of Hiroshima and Nagasaki (Kato and Schull, 1982). The four dose-response curves shown in the insert have been proposed (BEIR, 1980) to model the response from low doses of radiation: (*a*) linear, (*b*) general form, (*c*) linear quadratic, and (*d*) quadratic. Most radiologists support the use of the linear quadratic model (*c*).

(0.001 to 20 rad or 0.01 to 200 mGy). Other studies of exposures also in this lower range do not show an increase in leukemias above unexposed populations. Although above a dose of 6000 to 10,000 mrads (6 to 10 rads) the dose is linear, below that range identifiable radiation effects are not observed. The actual shape of the radiation exposure-effect curve at the low dose range may never be obtained due to the prohibitively large population sample size required for meaningful epidemiology to be performed. Meanwhile, radiation protection standards have followed the more conservative linear relationship (NCRP, 1987e). UNSCEAR (1986) has concluded that using a linear response model for estimating cancer risks when extrapolating from high dose rates could overestimate risks at low dose rates by a factor of five.

After nearly 40 years of studying the survivors of Hiroshima and Nagasaki, controversy remains over the doses received. These differences of opinion, however, are only in the two- to threefold range, and have a small effect on risk estimates for low-level exposures (Figure 6-4) (Roberts, 1987; UNSCEAR, 1988; BEIR, 1989). Recent reappraisals of the dose-response data may, however, have the effect

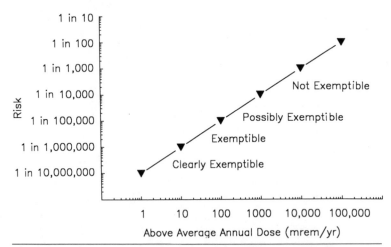

FIGURE 6-4. Relationship between exemptible risk and dose. Risks between one in a million and one in a thousand generally require analysis of the risk vs. cost benefit. Risks below that range are generally acceptable and those above it are not.

of lowering the current annual occupational dose limits. The current view is that the bomb used at Hiroshima delivered a much lower neutron dose and a two to three times higher gamma dose than had been previously thought, suggesting that, overall, the observed effects were caused by lower doses. The risk estimate for leukemia among the victims is now thought to have been two times greater, and for other cancers 50% greater, than among the general population. Even with therapeutic doses of ^{131}I that deliver an average whole-body dose of 10,000 mrems (10 mSv), nearly 22,000 patients treated for hyperthyroidism showed no increase in leukemia compared with surgical controls (Saenger et al., 1968). No health effects have been observed in people receiving doses below this level (BEIR, 1980). Without the need for models or extrapolations, studies of sizable human populations in Denver, Han (China), Kerala (India), and Araxa-Tapira (Brazil), where the background radiation levels are as much as ten times higher than average, on the order of thousands of mrem (mSv) per year, have not revealed higher incidences of cancer (Weinberg et al., 1987; Eisenbud, 1988).

The NRC has established maximum permissible limits for public exposure to manmade sources: 100 mrem (1 mSv) for continuous, 500 mrem (5 mSv) for infrequent, and an ALARA goal of 25 mrem (0.25 mSv) from a disposal site whole-body effective annual dose equivalent averaged over a lifetime (10 CFR Parts 20.105 and 61.41). They have assumed that an individual will not be exposed to more than four sources of 25 mrem (0.25 mSv) each and will not receive more than 100 mrem/yr (1 mSv/yr) total exposure. A 100-mrem (1-mSv) dose corresponds to a risk of mortality of approximately 1 in 100,000 per year (Figure 6-4) (BEIR, 1980; NCRP, 1987e). The correspondence of radiation dose levels to health and risk is shown in Figure 6-5. Since, on average, we all receive exposures of 364 mrem (3.64 mSv) annually from a combination of natural and man-made sources, additional exposures will increase this annual exposure by the amounts shown. It must also be considered that the exposures to high levels of radiation were generally acute, while exposures in the lower ranges were chronic. No clinical symptoms are observed below exposures of 10 rems (10,000 mrem, or 100 mSv); the maximum occupational

FIGURE 6-5. Relationship between dose, radiation effect, risk and mortality.

exposure is set at half that and the maximum continual exposure for the general public is 100 times less. The maximum dose limit for a radiation worker (5000 mrem) is at the same level of risk considered acceptable for U.S. industry (1:10,000). Risks of dying in automobile accidents or from cancer or heart disease are 2, 20, and 30 times higher, respectively. Below a risk of about 1:100,000, many activities and practices can be found whose risks are accepted by most individuals (NSC, 1988). The risk of a fatal cancer from the maximum permissible dose is ten times lower than the average annual risks accepted in most U.S. industries. At the lowest end of the exposure range, the NRC has defined *de minimis* risk as below 1 death in 10 million per year, corresponding to a dose equivalent of 1 mrem (0.01 mSv) (Kocher, 1986; 10 CFR Part 2). Put in perspective, this is equivalent to shortening one life expectancy by 1.2 minutes, making 4 street crossings, or riding 4 minutes in an automobile (Cohen and Lee, 1979).

Radiation is one of many environmental and man-made factors that may contribute, possibly synergistically, to the incidence of cancer. On the genetic level, radiation effects are qualitatively indistinguishable from the same effects with different causation (BEIR, 1980; Lloyd and Purrott, 1983). At the low radiation levels associated with LLRW, which are several times under background levels, radiation effects, if they do occur, are so infrequent that they would be masked by similar effects from other causes. Any health effects associated with *de minimis* levels are well below our limits of detection. The smallest risk, that is, excess number of cancer deaths, potentially observable without the sample size having to be prohibitively large (over 500,000 for control and exposed groups) would be approximately 3 in 10,000 (Alexander, 1988; Byrd and Lave, 1987). Figure 6-6 shows the relationship between dose and the size of the affected population needed to detect the effect. Since the annual risk for all cancers observed is 2 per 1000 persons, the risk is 0.002. If one additional cancer per 1 mrem (0.01 mSv) of radiation is predicted and that risk is 1 per million or 0.000001, then the total risk would be 0.002001. The "debate" among scientists is within a masked and narrow range, namely, whether the risk of cancer induction per additional rem (10 mSv) absorbed over a lifetime is 1, 2, or 3 in 10,000. It is therefore fruitless to conduct baseline epidemiological studies as part of an LLRW disposal siting process. Tumor registries already exist in most states. The demand for baseline studies will only result in further delays in the process, greater expense, and a reinforcement of the view that LLRW is exceedingly

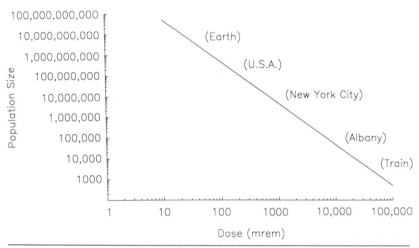

FIGURE 6-6. Population size required to observe effects from radiation exposures. Note that even for the highest doses, where the effects are more readily observed, the 1000 persons needed for the exposed group and 1000 persons for the control group must be observed for 25 to 50 years. The information given within parentheses is indicative of the size of the population required to study the effects of the corresponding dose. The range in which the majority of the potential exposures to LLRW lie (much below 100 mrem) requires prohibitively large, world-size populations.

hazardous. These delays may, however, be a political goal for forestalling LLRW disposal indefinitely.

Although 25% to 30% of the U.S. population will develop cancers of all origins, and 20% to 25% of the population is likely to die from cancer, the risk of dying prematurely of cancer from exposure to *de minimis* radiation is 100 times lower than the risk of death from suffocation (NSC, 1987). However, suffocation, unlike radiation, is not perceived by most of us as a significant cause of death. One personal reaction to risk is to consider the circumstances as either unlikely or avoidable, and that, to an extent, individual choices can be made. Reportedly, there is a 1 in 10,000 chance of a woman developing breast cancer from an X-ray; but there is a 1 in 11 chance she will develop breast cancer anyway, and the X-rays save many more lives if the cancer is detected early. However, individuals are not usually concerned with reducing the few mrem (a few hundredths of mSv) of radiation received at high elevations (as from working in a high-rise building, traveling by air, or living in Denver), from irradiation by bedrock when commuting by subway or working in certain buildings in New York City or sleeping next to someone. By simply sleeping beside someone else, one may receive an annual dose of up to 1 mrem (0.01 mSv) from the presence of naturally occurring isotopes in bone and muscle (Matuszek, 1988). Lastly, the possibility of hormetic effects, that exposure to low levels of radiation might even be beneficial, has also been recently discussed (Sagan, 1989; Wolff, 1989).

Above the *de minimis* range of 1 mrem (0.01 mSv) and below the MPD of 100 mrem (1 mSv), the scale on which most potential doses from LLRW can be measured, we should consider accepting BRC limits for public exposures so that our country's resources are not diverted to reducing risks to levels below which no detectable hazard or benefit can be realized.

Risk Management and LLRW

Except in a few specific cases, federal and state agencies subject disposal of licensed materials containing any level of radioactivity to regulation. Few *de minimis* levels have been recognized, generally applicable BRC levels have not been set, and, outside of a licensed site, the disposal of LLRW is dependent upon a specific application and permitting process. The maximum permissible concentrations of radionuclides in water and air, together with their projected impacts on a nearby population, govern these rulings (IAEA, 1987; 10 CFR Part 20). Prior to the temporary closure of the commercial disposal sites in the late 1970s, there was little incentive to seek disposal methods other than shallow land burial. The dominant and accepted conservative approach was to bury anything that had possibly been in contact with radioactive material as LLRW, regardless of the level.

In 1981, the NRC ruled that excreta from individuals undergoing medical diagnosis and treatment as well as aqueous wastes containing a limited amount of radioactivity [total of 5 Ci (185 GBq) ^{3}H, 1 Ci (37 GBq) ^{14}C, 1 Ci (37 GBq) other radionuclides per generator per year] could be disposed of as sewage (10 CFR Part 20.303). The NRC also ruled that animal carcasses and scintillation liquids containing less than 0.05 μCi/g (1.85 kBq/g) of ^{3}H or ^{14}C could be disposed of "without regard to its radioactivity" (10 CFR Part 20.306). This ruling suggested that incineration of these items would protect workers and the public from handling pathogenic and toxic chemical materials and, at the same time, reduce the volume and cost of a major fraction of institutional LLRW. The radioactivity released from the carcasses as water and carbon dioxide is so low as to only be detectable a few feet from the stack, providing doses on the order of 0.01 mrem (0.0001 mSv) to any individual in that vicinity, and contributing insignificantly to the natural background level (Roche-Farmer, 1980).

Formal recognition of *de minimis* levels for the common radioisotopes in all forms of waste is needed in order to address the various other hazards during the practical daily management of LLRW and to conserve disposal resources. Uniform and timely regulatory acceptance is also important, particularly because of the lag phase during which agencies with overlapping authority consider a new standard. For instance, the DOT, which controls the movement of NRC *de minimis* wastes, had a *de minimis* level of 0.002 μCi/g (0.074 kBq/g). Not until 1985 did they accept the 1981 NRC *de minimis* level of 0.05 μCi/g (1.85 kBq) ^{3}H and ^{14}C ruling (49 CFR Part 173). The EPA also waited until 1985 to accept this standard, and New York City's Health Department has still not accepted any level of radioactivity as *de minimis*.

In 1985, as part of the Amendments Act, Congress asked the NRC to make BRC rulings that would alleviate the LLRW disposal problem. In August 1986, claiming insufficient data to establish BRC rulings, the NRC published a guide (10 CFR Part 2) for petitioning to exempt specific waste streams from regulated disposal. On December 12, 1988, the NRC issued notification that they were proceeding with a generic exemption policy. In April 1989, they stated that this policy will no longer be referred to as BRC but as a policy "applicable to exemptions from some or all regulatory controls." Unlike the NRC's earlier *de minimis* ruling, these BRC decisions will become a matter of "compatibility" (compliance) for state as well as federal radiation protection programs. These rulings are likely to have a major impact on the LLRW problem, affecting both the volume and cost of disposal as well as the choice of the optimal disposal strategies.

At present we are spending as much money to avoid an unlikely and trivial exposure to LLRW as we are for the diagnoses and treatment of bona fide health problems. The cost for disposing of a 55-gal drum of LLRW is similar to the cost of spending a full day in intensive care. Our government is spending an estimated $200 million per eventual life saved to protect members of the public from the radioactivity in LLRW (Cohen, 1980). Compare this with the $25,000 spent on education and screening to prevent one death from cervical cancer or the $40,000 spent on smoke detectors to prevent one death (Eisenbud, 1988). The nuclear power industry assigns $25,000 to $50,000 per person-rem (hundredth of person-Sievert) of radiation exposure avoided. The DOE cost to avoid one statistical death is $5 million (Merkhofer et al., 1989). The unrealistic and irrational decisions determining LLRW disposal strategies currently being made throughout the nation threaten to drive these costs even higher. Although risk assessment is still in its infancy, results of these studies should be reevaluated in order to develop a more consistent national policy for the regulation of hazards.

Despite routinely basing risk assessments on "worst-case" scenarios, there is a question whether they are conservative enough (Bailar et al., 1989). On the other hand, many have argued that such estimates are too conservative. Perhaps, given that risk assessment is not precise, we should use the most probable, rather than the least likely, scenario on which to base our decisions. The latter approach would limit expenditures of huge sums of money that will not impact upon anyone's health, and those monies could be available for more socially important things now being neglected. The sudden furor in 1989 over the use of pesticides illustrates these points. According to the Natural Resources Defense Council, the EPA underestimated the cancer risk from residues of Alar in apple products 240-fold by using an outdated food consumption pattern for certain foods by preschoolers (Roberts, 1989). The EPA has announced that it will ban Alar, with which the study associated 86% to 96% of the cancer risk from pesticide residues. This cancer risk, however, was small. Of 22 million preschool children, 25%—or 5.5 million—are expected to get cancer during their lifetime anyway. The increase in cancers from pesticides translates to an increase from 25% to 25.025%. The likely result of this experience is that the use of many pesticides will be subject to stricter regulations. Ames and Gold (1989) emphasize that 99.99% of the pesticides we eat are naturally occurring and that the hypothetical cancer risk of 1 in a million or 1 in 100,000 should be balanced against the risks to our health and economy of not using these pesticides.

Societal Response to Risk Associated with Radiation

We have described LLRW and its different components and compared the low risks associated with it to other risks our society accepts. Unlike toxic chemicals and other hazardous wastes, much is known about exposure to LLRW. The benefits we derive from the materials and processes that generate LLRW far outweigh their risks. However, the common perception of these risks is just the opposite and not supported by the facts. As a result of this perception and despite the fact that the benefits of radioactive materials are widely shared socially and geographically, Congress has mandated that regional solutions to LLRW disposal be found. Neither the short- nor the long-term disposal of LLRW is a technical problem, yet disposal solutions are complicated by political and legal obstructions. In New York State the siting legislation is without scientific basis and contrary to the recommendations of the Governor's Advisory Committee (1984). This legislation has eliminated a geophysically sound site at West Valley and an established, cost-effective disposal strategy, shallow land burial, from consideration, while promoting engineered structures, an unjustifiably expensive strategy of uncertain performance. This legislative act has been, and will continue to be, costly to the community, the environment, and the workers at the site.

How have we responded to the LLRW problem? We have used political solutions to solve technical problems. These solutions are frequently superficial and dictated by the politician's view of the public's perception. Scientific opinion engenders three kinds of political responses, reflecting either agreement, disagreement, or uncertainty. In the first instance, some scientific opinion is disregarded in order to say there is agreement. Later, in the spirit of "compromise," scientific opinion is frequently augmented by other, nonscientific, opinions. They are not seen to overly complicate the issue, but they do. For example, in New York State scientists supported the legislation for establishing a LLRW burial site. However, the decision to exclude shallow land burial in the final legislation is unsupported by any scientific data. In the second case, the minority scientific opinion is represented as an equal, alternative point of view, evoking the response "the experts disagree; it is safer to err on the side of safety." Either of these first two approaches can result in overly restrictive regulations and inappropriate solutions. The last, and often favored, approach results in an interminable delay. To paraphrase a well-known answer of uncertainty to a preelection question is typical of this approach: "How do I stand on that issue? I am happy you asked me that question. Fifty percent of the experts are for it and fifty percent are against it. And I am one hundred percent for the experts." All three of these political responses have been observed within the LLRW issue. This is not to say political decisions are unnecessary, but they should not be cloaked in pseudoscience.

The goals of science and politics are quite different, arrived at through different processes. Political decisions address the wishes of the majority of the voting public or the needs of affluent, influential groups. Political decision making appears to require an all-or-nothing choice in which greater than 50% of the legislators are either for or against an issue. In contrast, science is a search for understanding in which all facts and opinions have to be counted and recounted. The investigative nature of the scientific approach does not lend itself to easily obtainable absolute conclusions. To scientists the political approach is a superficial and simplifying process. It is little wonder that scientists are frustrated by expedient political decisions that ignore some of the facts, while politicians are frustrated by the divergence of scientific opinion and the slow pace at which answers come forth. In general, our society is impatient with both politicians and scientists. We find it difficult to accept that what science provides is not necessarily the ultimate truth, but the best interpretation of the facts observed to date. Perhaps it is this element of uncertainty, which undermines our trust in experts, who, as more information becomes available, may change their opinions.

What then are reasonable expectations from a discussion of the "facts" related to LLRW? Unanimity of opinion may not always be forthcoming. Scientists may disagree on what evidence is "conclusive." However, with regard to exposures associated with LLRW, a consensus has been reached. Although risk assessment may not lead to precise values, acceptable ranges of radiation exposure have been recognized. Such consensus is essential for developing guidelines for any technology. Our governmental leaders need guidelines in order to avoid the high cost of basing compliance on conservative assumptions rather than on scientific evidence.

Minority opinions, although sometimes visionary, must be placed in perspective by experts and the media for the public. At a recent symposium on societal responses to LLRW, a health physicist stated that "what we heard from our scientific and technical speakers of exceptionally good quality, background, experience, and association this morning is . . . that the resistance to low level radioactive waste disposal sites is in fact a kind of political issue generated by environmentalists and people who don't understand the question, and really that the overall problem is, from a scientific and engineering view, a trivial one" (NYAM, 1988b). The last part of this statement was certainly a point upon which the participants agreed. However, he then stated that he disagreed with their view and proceeded to link the incineration of trash containing *de minimis* quantities of ^3H and ^{14}C with uncertainties about the safety of a new high-level-waste (HLW) repository. He further suggested that medical LLRW could be segregated from that of utilities and manufacturers since medical LLRW, "everybody agrees, is of favorable and desirable and relatively innocuous kinds of materials" (NYAM, 1988b). Yet, it was this same, innocuous medical LLRW that he had associated with HLW. The expert who reached this illogical conclusion also happened to be a health department official. His opinion is therefore of greater political influence.

Not only are there too many experts and persons in positions of authority who communicate poorly or inappropriately, but there are also many more who make little effort at all. We need experts and leaders who are informed and able to communicate. But even if we had them, it is the media to whom we turn for information. A National Cancer Institute survey (1984) indicated that 64% of the U.S. population obtains information on cancer prevention from magazines, newspapers, and television, while only 13% to 15% request this information from their physicians. Holding the media alone to blame seems too simplistic. The media do not tell us what to think, but rather what to think about (Nelkin, 1989). Their decision on what is newsworthy tends to identify policy issues. The metaphors used by the media to characterize a news item as an "accident," "incident," "disaster," or an "event" can greatly influence the thrust of those policies. "Deadly" has been subliminally linked to radiation and "dump" has become synonymous with a disposal facility. Their use can affect both risk assessment and communication as the following headline shows.

Nuclear **Dump** Plan Ignites Rural Protests

Webster's Ninth New Collegiate Dictionary provides the following definitions of "dump" from which the headline gains its negative connotations.

vt	1:	to let fall in a heap or mass
	2:	to get rid of unceremoniously or irresponsibly
n	1:	an accumulation of refuse and discarded materials
	2:	a disorderly, slovenly. or dilapidated place
adj	1:	shabby, dingy

We should have initiated grammar school programs to provide technical information and understanding to the nonscientifically inclined 30 years ago. At this stage various efforts are warranted but they may come too late for the waste crisis. Moreover, supplying technical information frequently provokes a negative public reaction. Even when done well, for many it will be too difficult to assimilate due to lack of background or more often lack of interest. Likewise, quantities, so essential to scientific evaluation and risk assessment, are often not meaningful to the majority of people who are now becoming aware that many agents can be hazardous in small amounts.

Risk communicators stress the importance of open, two-way communication instead of public relations (Covello, 1989; Krimsky and Plough, 1988), but regardless of the quality of communication, do we treat risks that are quantitative the same way we treat risks that are judgmental? The former are of the type we inherently distrust and the latter are heavily influenced by other factors. Although experts tend to dismiss community judgments as irrational, these judgments may in fact appear rational when based on different experiences, values, and standards of evidence. For example, learning that several neighbors in a small community have cancer while simultaneously discovering the presence of hazardous wastes in that community can lead to unsubstantiated assumptions of cause and effect.

In the effort to locate LLRW sites, many states across the country have engaged in extensive education programs, marked by public meetings, media events, and publication. In general, the public response has been characterized by disinterest, distrust, and concern for property values (Kobrinsky, 1989). Frequently, receptiveness, if not responsiveness, is governed by the degree to which individuals feel personally involved. These statewide efforts have been accompanied overall by very small public response over the 2-year preliminary period during which surveys of potential locations were taken. After thousands of announcements for public meetings to discuss New York State plans for LLRW disposal were mailed, only two dozen people attended in New York City and only a few were not involved in the program. In contrast, announcements of potential site locations bring out a visible and verbal public who feel frustrated over the small opportunity they will now have to avert the location of a waste site in their backyard. Three commissioners on their way to a public meeting in New York to provide information on a potential site were trapped in their cars by an angry crowd that was dispersed only with police assistance. Some years ago, intelligent and interested individuals in a state office of science and technology eagerly absorbed the details of the unfolding LLRW story. With the expectation that they would attempt to avert the LLRW disposal crisis or develop contingency plans, they were asked what their next course of action would be. They replied "the information would be very helpful to have on file in order to shorten the downtime when the crisis occurred." Confronted by crisis, people usually do get involved, creating a reactionary system that is out of synchrony with our dynamic life-style.

There are lessons to be learned from the LLRW disposal siting process. In the short term, the only success in gaining community acceptance has been to find remote areas with small, economically depressed populations and address the economic impact on the selected communities. Concern for property values will have to be offset by financial incentives in the form of jobs, scholarships, and direct compensation. In the long term, not much will change in terms of societal understanding or acceptance of risk. A relatively small group of people will still feel they have been unjustly singled out to bear the burden of any waste disposal site. The vast majority will retain their prejudices and conviction that the solution is sound as long as it is "not in my backyard."

The obvious answer, especially to readers with strong academic backgrounds, might be education. Of course, education on the fundamentals of risk, radiation, toxic chemicals, and waste, beginning in grammar school and continuing through college, may eventually contribute to improving the general level of understanding in this area. However, altering life-styles and opinions through education is a slow and incomplete process as evidenced by efforts to control smoking, drinking, driving, or wearing seat belts. For driver education, efforts at both the high school and adult driver levels have been shown to be counterproductive (Robertson, 1986, 1988). Behavior modification requires our constant attention. Seat-belt use has depended upon maintaining costly saturation advertising (NYAM, 1988a). From attendees at seminars specifically designed to teach laypersons about nuclear waste, the League of Women Voters Educational Fund (1989) concluded that "those who came with a definite opinion left with the same opinion, and those who came with an open mind, undecided about nuclear wastes issues, left still undecided, though some said that the seminar had given them the information they needed to make up their minds." For scientists as well as other academics and educators, the failure of our educational approach is difficult to accept. It must be considered, however, that many in the United States do not receive such formal training and the quality of what they do receive is uneven. In New York City the high school dropout rate is close to

30%; 25% of those who do graduate have repeated at least 1 year in high school. Although the rate of functional literacy may be low in the United States, scientific literacy here has recently been rated as low as 5% (Massey, 1989). It should not be surprising that much of what we "learn" as facts and accept as "logical conclusions" is the product of many factors other than formal education. Key among them is the opinion of our parents, peers, and social contacts. The broadcast news is our society's major source of information, but it tends to reinforce existing views. We remain unwilling to listen to or accept information that may challenge concepts basic to our life-styles, and are therefore inconsistent in our acceptance of risk; for example, we accept the high risk of automobile fatality but are unwilling to accept the low risk of cancer from LLRW.

We are more interested in the benefits of technology than in the knowledge that supports it. We do not want to know how it works, only how to turn it on. We expect our scientists to collect data, clearly state facts, and derive logical conclusions. Yet, simply stating the facts does not necessarily lead to logical or even anticipated conclusions. This may explain the apparent low level of participation by many scientists and poor performance by others who try to communicate technical information and risk. We marvel at technical achievements despite the fact that technical details are often unwanted, overwhelming, and if they alter opinion, rejected. Our society appears to be fomenting and formalizing confrontation with technology and its developers. The relatively new and aggressively pervasive concept of "right-to-know" and the basic rule of "informed consent" have become mainstays of our technocracy. Although education, risk communication, and the other factors discussed are important contributors to our perception of the risks associated with low-level radiation, it is our behavior, not our understanding, that must be modified. Such societal changes are not unachievable, but they are slow in relation to the expanding array of technological and environmental challenges.

One has only to compare the standard of living and life expectancy in the industrialized and underdeveloped nations to see the benefits of a technological society. We cannot wait to use hindsight to deal with the waste by-products of today's and tomorrow's technology. We should place more faith in our technical abilities to solve our waste problem instead of giving up the many benefits associated with it because we are afraid to deal with our wastes. If we will not understand risk issues or entrust our waste disposal decisions to those who can, we will have to pay for overdesigned disposal facilities and for communities to host them in their backyards. "We should make decisions on the basis of what we know, not on the basis of what we do not know, have not yet learned, or may never learn" (Lutzker, 1989).

References

Alexander, R. 1988. An appalling place to start. *Health Physics Society's Newsletter* 16(3).

Ames, B. N., and L. S. Gold. 1989. Pesticides, risk, and applesauce. *Science* 244:755–757.

Ames, B. N., R. Magaw, and L. S. Gold. 1987. Ranking possible carcinogenic hazards. *Science* 236:271–280.

Auxier, J. A., and H. S. Dickson. 1983. Guest editorial: Concern over recent use of the ALARA philosophy. *Health Physics* 44:595–600.

Bailar, J. C. Crouch, R. Shaikh, and D. Spiegelman. 1988. One-hit model of carcinogenesis: conservative or not? *Risk Analysis* 8:485–497.

Berg, G. G., and H. D. Maillie. 1981. *Measurement of Risks*. New York: Plenum Press.

Boice, J. D., Jr., and J. F. Fraumeni, Jr. 1984. *Radiation Carcinogenesis Epidemiology and Biological Significance, Progress in Cancer Research and Therapy*, Vol. 26. New York: Raven Press.

Byrd, D., and L. B. Lave. 1987. Narrowing the range: A framework for risk regulators. *Issues in Science and Technology* Summer:92–100.

Cohen, B. 1980. Society's valuation of life saving in radiation protection and other contexts. *Health Physics* 38:33–51.

Cohen, B., and I. S. Lee. 1979. A catalog of risks. *Health Physics* 36:707.

Comar, C. 1979. Risk: A pragmatic *de minimis* approach. *Science* 203:319.

Committee on the Biological Effects of Ionizing Radiations (BEIR). 1980. *The Effects on Populations of Exposure to Low Levels of Ionizing Radiation: 1980*. Washington, D.C.: National Academy Press.

Committee on the Biological Effects of Ionizing Radiations (BEIR). 1988. *Health Risks of Radon and Other Internally Deposited Alpha-Emitters: BEIR IV*. Washington, D.C.: National Academy Press.

Committee on the Biological Effects of Ionizing Radiations (BEIR). 1989. *Health Effects of Exposure to Low Levels of Ionizing Radiation: BEIR V*. Washington, D.C.: National Academy Press.

Comptroller General of the United States. 1981. *Problems in Assessing the Cancer Risks of Low-Level Ionizing Radiation Exposure*. Report to Congress, Vols. 1 & 2. Gaithersburg, Maryland: U.S. General Accounting Office.

Covello, V. T. 1987. Decision analysis and risk management decision making: Issues and methods. *Risk Analysis* 7(2):131–139.

Covello, V. T. 1989. Symposium on Science and Society: Low Level Radioactive Waste. Controversy and Resolution. *Bulletin of the New York Academy of Medicine* 65:467–482.

Eisenbud, M. 1987. *Environmental Radioactivity*, 3d. ed. Orlando, Fla.: Academic Press.

Eisenbud, M. 1988. Disparate costs of risk avoidance. *Science* 241:1277.

Eisenbud, M. 1988. Low-level radioactive waste repositories: A risk assessment. *Annual Conference of the North Carolina Academy of Science*,
March 26, 1988. Charlotte, N.C.

Federal Register. 1989. 54:6392.

Federal Reporter. 1987. *Public Citizen v. Young.* Cite as 831 F.2d 1108 (D.C.
Cir. 1987).

Governor's Advisory Committee. 1984. *New York State Energy Office Low-
Level Radioactive Waste Management Study*, Vols. 1–3. Albany, N.Y.:
New York State Energy Office.

Houk, V. 1989. Symposium on Science and Society: Low Level Radioactive
Waste. Controversy and Resolution. *Bulletin of the New York Academy
of Medicine* 65:485.

Gruhlke, J. M., F. L. Galpin, W. F. Holcomb, and M. S. Bandrowski. 1989.
USEPA's proposed environmental standards for the management and land
disposal of LLRW and NARM Waste. *Waste Management '89* 2:273–276.

Holcomb, W. F., J. M. Gruhlke, and F. L. Galpin. 1989. USEPA's proposed
standard for BRC criteria. *Waste Management '89* 2:361–364.

International Atomic Energy Agency. 1987. *Exemption of Radiation Sources
and Practices from Regulatory Control: Interim Report.* IAEA-TECDOC-
401. Vienna, Austria: IAEA.

Jacobs, D. G. 1968. *Sources of Tritium and Its Behavior upon Release to the
Environment.* Oak Ridge, Tenn.: AEC.

Kato, H., and W. J. Schull. 1982. Studies of the mortality of A-bomb survivors. Mortality, 1950–1978: Part I. Cancer Mortality. *Radiation Research* 90:395–432.

Kobrinsky, C. 1989. Symposium on Science and Society: Low Level Radioactive Waste. Controversy and Resolution. *Bulletin of the New York Academy of Medicine.* 65:527–531.

Kocher, D. C. 1986. A proposal for a generally applicable *de minimis* dose.
Health Physics 53(2):117–121.

Krimsky, S., and A. Plough. 1988. *Environmental Hazards: Communicating
Risks as a Social Process.* Dover, Mass.: Auburn House.

Kumazawa, S., D. R. Nelson, and A. C. Richardson. 1984. *Occupational
Exposure to Ionizing Radiation in the United States: A Comprehensive
Review for the Year 1980 and Summary of Trends for the Years 1960–
1985.* Report EPA 520/1-84-005. Washington, D.C.: EPA.

Lave, L. B. 1987. Health and safety risk analyses: Information for better
decisions. *Science* 236:291–295.

League of Women Voters Educational Fund 1989. *Nuclear Waste Education
Project, Final Report Albuquerque Seminar.* Washington, D.C.: League
of Women Voters' Educational Fund.

Lloyd, D. C., and R. J. Purrott. 1983. *The Study of Chromosome Aberration Yield in Human Lymphocytes as an Indicator of Radiation Dose: Revised Techniques.* Report NRPB-M70. Chilton, England: NRPB.

Lowrance, W. W. 1983. The agenda for risk decisionmaking. *Environment* 25(10):4–8.

Lutzker, L. 1989. Symposium on Science and Society: Low Level Radioactive Waste. Controversy and Resolution. *Bulletin of the New York Academy of Medicine.* 65:495.

Massey, W. E. 1989. Science education in the United States: What the scientific community can do. *Science* 245:915–921.

Matanoski, G. M., R. Seltser, P. E. Sartwell, E. L. Diamond, and E. A. Elliot. 1975. The current mortality rates of radiologists and other physician specialists: Specific causes of death. *American Journal of Epidemiology* 101(3):199–210.

Matuszek, J. M. 1988. Safer than sleeping with your spouse—The West Valley experience. In *Low-Level Radioactive Waste Regulation: Science, Politics, and Fear,* ed. M. E. Burns, Chelsea, Mich.: Lewis Publishers, pp. 260–278.

Merkhofer, M. W., T. A. Cotton, K. E. Jenni, J. C. Lehr, and T. P. Lonjo. 1989. A program optimization system for the cleanup of DOE hazardous waste sites, and application to FY 1990 funding decisions. *Waste Management '89* 2:149–158.

Mumpower, J. 1986. An analysis of the *de minimis* strategy for risk management, *Risk Analysis* 6(4):437–446.

National Academy of Sciences (NAS) 1980. *Disposal of Low-Level Radioactive Biomedical Wastes.* Washington, D.C.: National Academy Press.

National Cancer Institute 1984. *Cancer Prevention Awareness Survey.* NIH 84-26-77. Washington, D.C.: U.S. Government Printing Office.

National Council on Radiation Protection and Measurements. (NCRP). 1976. *Environmental Radiation Measurements.* NCRP Report No. 50. Bethesda, Md.: NCRP.

National Council on Radiation Protection and Measurements (NCRP). 1977. *Cesium-137 from the Environment to Man: Metabolism and Dose.* NCRP Report No. 52. Bethesda, Md.: NCRP.

National Council on Radiation Protection and Measurements (NCRP). 1979. *Tritium in the Environment.* NCRP Report No. 62. Bethesda, Md.: NCRP.

National Council on Radiation Protection and Measurements (NCRP). 1980. *Perceptions of Risk.* NCRP Proceedings No. 1. Washington, D.C.: NCRP.

National Council on Radiation Protection and Measurements (NCRP). 1981. *Quantitative Risk in Standards Setting.* NCRP Proceedings No. 2. Bethesda, Md.: NCRP.

National Council on Radiation Protection and Measurements (NCRP). 1982. *Critical Issues in Setting Radiation Dose Limits.* NCRP Proceedings No. 3. Bethesda, Md.: NCRP.

National Council on Radiation Protection and Measurements (NCRP). 1983. *Environmental Radioactivity.* NCRP Proceedings No. 5. Bethesda, Md.: NCRP.

National Council on Radiation Protection and Measurements (NCRP). 1983. *Iodine-129: Evaluation of Releases from Nuclear Power Generation.* NCRP Report No. 75. Bethesda, Md.: NCRP.

National Council on Radiation Protection and Measurements (NCRP). 1984. *Radiological Assessment: Predicting the Transport, Bioaccumulation, and Uptake by Man of Radionuclides Released to the Environment.* NCRP Report No. 76. Bethesda, Md.: NCRP.

National Council on Radiation Protection and Measurements (NCRP). 1985a. *Carbon-14 in the Environment.* NCRP Report No. 81. Bethesda, Md.: NCRP.

National Council on Radiation Protection and Measurements (NCRP). 1985b. *Induction of Thyroid Cancer by Ionizing Radiation.* NCRP Report No. 80. Bethesda, Md.: NCRP.

National Council on Radiation Protection and Measurements (NCRP). 1986. *Radioactive Waste.* NCRP Proceedings No. 7. Bethesda, Md.: NCRP.

National Council on Radiation Protection and Measurements (NCRP). 1987a. *Exposure of the Population in the United States and Canada from Natural Background Radiation.* NCRP Report No. 94. Bethesda, Md.: NCRP.

National Council on Radiation Protection and Measurements (NCRP). 1987b. *Ionizing Radiation Exposure of the Population of the United States.* NCRP Report No. 93. Bethesda, Md.: NCRP.

National Council on Radiation Protection and Measurements (NCRP). 1987c. *Public Radiation Exposure from Nuclear Power Generation in the United States.* NCRP Report No. 92. Bethesda, Md.: NCRP.

National Council on Radiation Protection and Measurements (NCRP). 1987d. *Radiation Exposure of the U.S. Population from Consumer Products and Miscellaneous Sources.* NCRP Report No. 95. Bethesda, Md.: NCRP.

National Council on Radiation Protection and Measurements (NCRP). 1987e. *Recommendations on Limits for Exposure to Ionizing Radiation.* NCRP Report No. 91. Bethesda, Md.: NCRP.

National Low-Level Radioactive Waste Management Program (NLLRWMP). 1980. *The 1979 State-by-State Assessment of Low-Level Radioactive Wastes Received at Commercial Disposal Sites.* NUS-3340. Springfield, Va.: NTIS.

National Safety Council (NSC) 1988. *Accident Facts.* Chicago, Ill.: NSC.

Nelkin, D. 1989. Communicating technological risk: the social construction of risk perception. *Annual Review of Public Health* 10:95–113.

New York Academy of Medicine (NYAM) 1988a. Symposium on Motor Vehicle Injuries. *Bulletin of the New York Academy of Medicine* 64:605–866.

New York Academy of Medicine (NYAM) 1988b. Symposium on Science and Society: Low Level Radioactive Waste. Controversy and Resolution. *Bulletin of the New York Academy of Medicine* 65:483–554.

Okrent, D. 1987. The safety goals of the U.S. Nuclear Regulatory Commission. *Science* 236:296–300.

Perera, F. 1985. New approaches in risk assessment for carcinogens. *Risk Analysis* 6(2):195–201.

Riggan, W. B., J. Van Brugger, J. F. Acquavella, J. Beaubier, and T. J. Mason 1983. *U.S. Cancer Mortality Rates and Trends 1950–1979.* EPA-600/1-83-015a. Washington, D.C.: NCI/EPA.

Roberts, L. 1987. Atomic bomb doses reassessed. *Science* 238:1649–1651.

Roberts, L. 1989. Pesticides and kids. *Science* 243:1280–1281.

Robertson, L. S. 1986. Behavioral and environmental interventions for reducing motor vehicle trauma. *Annual Review of Public Health* 7:13–34.

Robertson, L. S. 1988. Driver education: The mix of science and ideology. *Bulletin of the New York Academy of Medicine* 64(7):617–622.

Roche-Farmer, L. 1980. *Study of Alternative Methods for the Management of Liquid Scintillation Counting Wastes.* NUREG/CR-0656. Washington, D.C.: NRC.

Rogers, V. C., and E. S. Murphy. 1987. *Disposal of Short-Lived Radionuclide Wastes in a Sanitary Landfill.* Austin, Tex.: Texas Low-Level Radioactive Waste Disposal Authority.

Russell, M., and M. Gruber. 1987. Risk assessment in environmental policy-making. *Science* 236:286–290.

Saenger, E. L., G. E. Thoma, and E. A. Thompkins. 1968. Incidence of leukemia following treatment of hyperthyroidism. *Journal of the American Medical Association* 205:855–862.

Sagan, L. A. 1989. On radiation, paradigms, and hormesis. *Science* 245:574.

Schiager, K. J., W. J. Bair, M. W. Carter, A. P. Hull, and J. E. Till. 1986. *De minimis* environmental radiation levels: Concepts and consequences. *Health Physics* 50(5):569–579.

Shapiro, J. 1981. *Radiation Protection: A Guide for Scientists and Physicians,* 2d ed. Cambridge, Mass.: Harvard University Press.

Silverberg, E., and J. Lubera. 1986. Cancer statistics, 1986. *Ca-A Cancer Journal for Clinicians.* 36:9-25.

Slovic, P. 1987. Perceptions of risk. *Science* 236:280–285.

Supreme Court of the United States. 1980. *Industrial Union Department, AFL-CIO v. American Petroleum Institute.* 448 U.S. 607.

Tichler, J., and K. Norden. 1986. *Radioactive Materials Released from Nuclear Power Plants—Annual Report 1983.* NUREG/CR-2907, Vol. 4. Washington, D.C.: NRC.

Tirmache, M. 1988. *Carcinogenic Risk for Workers Exposed to Ionizing Radiation. A Critical Review of Present Epidemiologic Studies.* Epidemiology and Radiation Protection, Paris, France: OECD.

United Nations Scientific Committee on the Effects of Atomic Radiation (UNSCEAR). 1986. *Genetic and Somatic Effects of Ionizing Radiation.* Report to the General Assembly. New York: U.N.

Weinberg, C. R., K. G. Brown, and D. G. Hoel. 1987. Altitude, radiation, and mortality from cancer and heart disease. *Radiation Research* 112:381–390.

Wilson, R., and E. A. C. Crouch. 1987. Risk assessment and comparisons: An introduction. *Science* 236:267–270.

Wolff, S. 1989. Are radiation-induced effects hormetic? *Science* 245:575.

Yalow, R. 1988a. Biologic effects of low-level radiation. In *Low-Level Radioactive Waste Regulation: Science, Politics, and Fear*, ed. M. E. Burns, Chelsea, Mich.: Lewis Publishers, pp. 239–259.

Yalow, R. 1988b. The International Symposium on Nuclear Medicine, Beijing, People's Republic of China, 10–14 October 1988.

Yalow, R. 1989. The contributions of medical physicists to radiation phobia. *Medical Physics* 16:159–161.

7 CONCLUSIONS

Although LLRW constitutes less than 1% of the nonmunicipal waste in the United States, resolution of its proper disposal is essential if we are to meet the greater environmental challenges from the hazardous wastes that we continue to produce in ever increasing amounts. While far from perfect, better environmental control can be obtained at the three isolated and well-characterized remaining commercial LLRW disposal sites than from toxic chemical sites. Radionuclides migrating from the sites are easy to detect and analyze and their potential health effects are fairly well established. The great diversity of chemical contaminants at hazardous waste sites makes their detection and analysis both difficult and costly. Furthermore, the complete absence of disposal records for many chemical sites is accompanied by a lack of toxicological data and epidemiological study on most of the chemicals. Given that the inventories at LLRW sites are much better documented, the nature and rates of chemical as well as radionuclide migration from these sites could be studied as possible models for migration at existing toxic chemical waste sites. The opportunity to solve and learn from the LLRW disposal problem may not come again before widespread economic and environmental damage is done.

Prior to the temporary closure of the commercial disposal sites in the late 1970s, there was little incentive for the United States to seek alternative methods to shallow land burial. The dominant and accepted approach was to bury any materials that had been in possible contact with radioactive material, regardless of the level of contamination. As in Europe, the United States has finally recognized its finite capacity for land disposal and its associated liabilities and has begun to focus on volume-reduction and waste minimization. As evidenced by the success of some industries, institutions, and nuclear power utilities, the segregation and exclusion of short-lived isotopes and deregulation of a broader range of materials, combined with waste processing, can reduce the volume of LLRW requiring burial by at least 80%. Although sections of the Amendments Act require nuclear power plants to institute major volume-reduction measures, similar requirements for institutional and industrial generators do not exist. Simply because institutional and industrial LLRW account for smaller fractions of all LLRW does not diminish the importance of waste minimization for these generators. Reducing the large volume of low-activity Class A material that they generate will help avoid the historical problems associated with the management of shallow land disposal (i.e., trench subsidence, water infiltration, and subsequent isotope migration). In addition to decreased costs, volume reduction extends the expected life of any disposal facility, reduces the number of people exposed to LLRW, and allows for more time and energy to be focused on the small amount of higher-activity and longer-lived wastes that remain. This should influence both the selection of disposal site(s) and disposal methods.

Volume-reduction methods and management controls are the keys to safe and efficient waste disposal. Because some of the technology is expensive and difficult for a small generator to accommodate logistically, services should be provided at treatment centers or at disposal sites. Decay centers could be provided for generators who cannot accommodate a decay program on-site. Incineration is efficient and cost-effective and can be applied to approximately 70% of LLRW. Emissions can be controlled and monitored, the nonradiologic hazards destroyed, and the volume reduced to a more manageable form. This treatment provides the best and least expensive approach to long-term management. Supercompaction and vitrification, if the latter can be made cost-effective, can also be used to further increase the density and provide greater stabilization of LLRW. Regulatory agencies should complement the efforts of the generators by deregulating those levels of radioactivity that do not pose significant health risks and work to coordinate disposal practices. By a combination of these processes and exclusion of BRC and *de minimis* materials, overall reduction of LLRW could reach 95%. Volume-reduction measures can be anticipated to increase the concentration of radioactivity of the remaining wastes by several orders of magnitude. Although not profitable for the site operators, volume-reduction measures are essential for proper site management.

All of the land-based disposal methods now regulated by the NRC under 10 CFR Part 61 are adequate to protect our health and environment, but their success depends on proper waste management. While each method differs in its dose potential to site operators, all minimize exposures to the public. Although the history of the commercial burial sites in the United States indicates that shallow land burial (SLB) can be successfully practiced, even without the present rigorous siting criteria and management goals, most compacts and independent states are studying above- or below-ground vaults. The capital costs for these methods will be much higher than for SLB.

Considering that exposures to the public from SLB have been so low, such added expense is not justified. Above-ground vaults are more susceptible to intrusion and environmental factors, and, lacking surface barriers, leakage from them will spread more rapidly than from other types of facilities and may actually exceed regulatory limits. Insistence upon costly engineered above- or below-ground structures that require extensive monitoring and maintenance schemes has made establishing a facility less feasible and undermined the ability of the states to successfully formulate disposal solutions. Our insistence upon these costly, overengineered facilities underscores our misunderstanding of LLRW and its risks. Part of this lack of understanding has led to a political approach that suggests that wastes from different, that is, institutional, sources be segregated and then treated independently. Despite the differences between the nature of the source and composition, generators produce LLRW of remarkably similar form and activity. Furthermore, industrial and institutional generators are intrinsically linked because a large percentage of industrial LLRW results from producing radiochemicals for clinical and research applications. Solutions appropriate for most utility LLRW, which is the majority of all commercial LLRW, will also be appropriate for the wastes from other generators.

The superficial political approaches to the United States' LLRW disposal problem reflect the polarization of our society on issues involving radiation. A major obstacle to our thinking is our understandable preoccupation with nuclear holocaust. It is difficult to believe that many people think of war, much less a nuclear war, as an acceptable concept, but a sufficient number consider the maintenance of nuclear arsenals to be a viable, even necessary, deterrent to world war. Unfortunately, the fallout from this approach is a large part of what divides our society. Lack of technical understanding has led us to think that being in favor of the use of radioactivity for any purpose is supporting nuclear weapons and that we must be either pro- or antinuclear. Nothing could be further from the truth. Yet, LLRW may well be the fulcrum on which all the beneficial applications of man-made radiation are balanced. A publication from one antinuclear group supports "an orderly, rational phase-out of nuclear reactors. While . . . underway, production of nuclear wastes must be minimized" (Resnikoff, 1987). Because many of us have failed to understand that the health and status of our society depend upon the use of radioactive materials, there is a danger that "minimization" will be interpreted to mean elimination before we have developed alternative technologies to supply our societal needs. We must recognize that most of the nationally shared benefits from the use of radioactive materials far outweigh the small risks associated with the wastes they produce. In the United States we are now spending as much money to avoid an unlikely and trivial exposure from LLRW as we are for the diagnoses and treatment of health problems. The cost for disposing of a 55-gal drum of LLRW is similar to the cost of spending a full day in intensive care; we spend an estimated $200 million per eventual life saved to protect us from the radioactivity in LLRW (Cohen, 1980). The mandates of the Amendments Act and the LLRW disposal strategies proposed to meet those mandates threaten to drive these costs even higher. The threat is so real that medical procedures and research using radioactive materials may stop if states fail to develop disposal options, or if they choose prohibitively expensive strategies. Serious health penalties may arise from our emotional reaction to trivial levels of radioactivity in LLRW. We must reinstill the idea that science and technology are sources of benefits rather than causes for fear and concern.

Because the U.S. government has more resources and experience (albeit much of it negative) than any state in the management of LLRW, its decision to pass the responsibility for locating additional disposal sites to the states was inappropriate and irresponsible. The benefits of radionuclide manufacture and use are nationwide. Transportation of LLRW has been a coast-to-coast exercise for many years without evidence of serious mishap. Siting new LLRW disposal facilities is very difficult and costly for regional compacts. The few volume-reduction measures the federal government has legislated have been effective. In fact, the reduction in volume to date has made it less lucrative for the three commercial disposal sites to operate, and has resulted in higher burial costs. The establishment of more sites, possibly 16 in all, will make it prohibitively expensive for any of them to operate, provide disincentives for waste minimization and volume reduction, and have a negative influence on generator practices. It would be far more cost effective for the federal government to institute more extensive waste- and volume-reduction measures and help establish a national site. National or at least centralized waste processing and disposal has been the approach taken by other industrialized nations. These problems and the tentative nature of regional compacts make it likely that the federal government will still need to intervene to manage LLRW after 1993.

One of the few benefits of the regional siting process is that different areas of the country have been highly scrutinized by different groups well versed in the history, applicability, and feasibility of these areas to serve as waste disposal sites. The result will be the identification of a dozen or so potential waste disposal sites. These sites could be compared and the most appropriate selected as a national LLRW disposal site; the remainder could be considered for other kinds of waste treatment and disposal facilities. Not-in-my-backyard responses are inevitable, but unless we confront and constructively respond to our waste crises, there will be no viable backyards.

The United States should initiate both a long-term and a short-term strategy for locating waste disposal sites. The long-term approach to waste disposal problems entails altering our attitudes and understanding of wastes. We must improve waste-oriented technical education and communication of risk assessment throughout our schooling. This effort is most important for the majority of people who are not particularly interested in science. These efforts must be championed by informed public officials and educators using guidelines built on the rigors of science with the support of the media. The development of technical guidelines depends upon waste categories that encompass all forms and properties, upon accurate and readily available data on waste generation and disposal, and valid risk assessment. Unfortunately, the only short-term approach upon which we can immediately rely is to offer financial incentives to geophysically compatible, economically depressed areas in exchange for accepting waste disposal sites. Would it not make more sense to provide incentives for one or at most a few national sites in the most suitable locations in the country rather than to dilute our efforts to include less favorable and unneeded locations?

Although disposal solutions for LLRW are technically relatively simple, the United States' political responses to the disposal problem have placed complex legislative, economic, and logistical obstacles in the path of its resolution. If we are to conserve many of the benefits now characteristic of our life-style, we must more clearly define an effective course of action for waste disposal. Although the disposal of LLRW is being addressed, analyzing the process, both past and present, will provide insights into the waste crisis in the industrialized world in general. The time to reassess is now, but the time to act is running out.

It is with some irony that the eighteenth-century English economist, Malthus, focuses his concerns about population growth on the input, rather than the output, side of the supply chain. Namely, that populations would continue to expand to the level of subsistence unhindered except by famine, war, and ill health. Perhaps this reflects a basic difference between the overriding concern of eighteenth-century persons and, what is now safe to say, will be a twenty-first-century point of view. Malthus did not foresee the benefits, or the by-products, of the industrial revolution and the technologies it inspired. Although malnutrition remains a major concern, especially in less industrialized nations, it is unlikely that it will be a lack of food or the exhaustion of raw materials that halts the growth of humankind, but rather it will be the waste by-products of our industrialized societies.

Economic growth, usually expressed as gross national product (GNP) per capita, correlates well with both energy consumption and waste production. Even the more experienced among the industrialized nations are still struggling to formulate waste disposal strategies to treat the municipal garbage and sewer sludge they produce. Pollution in our environment is really a matter of concentration, the amount of toxic material per unit of living space. With finite living space, the population's size and ability to pollute ultimately determines concentration. The less experienced, more rapidly growing populations will industrialize at a more rapid rate than their predecessors— a forecast for a dramatic increase in consumption and waste production. They will of necessity be more concerned with the products, not the by-products, of their technological achievements. Thus the time for solving waste problems will vary inversely with the rate of increase in energy consumption and GNP per capita.

Malthus recognized that in order to maximize wealth, a nation must balance its power to produce with its will to consume. Our present recognition of the role we have played on our planet is derived from Rachel Carson's *Silent Spring* and other world-view-oriented publications of the 1960s and 1970s. Our sense that those who generate the waste should be responsible for its disposal has been translated into the RCRA and Superfund legislation and is defined in the concept "from cradle to grave." However, not all generators are aware of the potential hazards of the wastes they discard. Even if they are aware and accepting of the responsibility for their disposal, a mechanism for proper waste treatment and disposal must be available in order to fully discharge that responsibility. Frequently, the availability of an appropriate means of disposal depends on the scale of waste produced. In the United States each of us will generate an estimated one ton of waste per year. Commercial operations may produce many-fold more. RCRA-regulated hazardous waste is reportedly 10 to 50% of this volume. Although we have the technology for disposal of hazardous waste, we have not yet provided for, or in many instances allowed, the proper treatment of hazardous wastes. This is at the center of the waste disposal crisis. As a microcosm, the LLRW story is an important introduction to this larger crisis.

References

Cohen, B. 1980. Society's valuation of life saving in radiaton protection and other contexts. *Health Physics* 38:33–51.

Resnikoff, M. 1987. *Living Without Landfills*. New York: Radioactive Waste Campaign.

APPENDICES

Additional Reading

Introduction to Radiation

Cobb, Charles, E., Jr. 1989. Living with Radiation. *National Geographic.* April.

Hall, Eric J. 1984. *Radiation and Life.* New York: Pergamon Press. ISBN 0-080288-19-7.

Lillie, David W. 1986. *Our Radiant World.* Ames, Iowa: Iowa State University Press. ISBN 0-8138-1296-8.

Wagner, Henry N., Jr., and Linda E. Ketchum. 1989. *Living with Radiation.* Baltimore, Md.: The Johns Hopkins University Press. ISBN 0-8018-3787-1.

Health Physics

Cember, Herman. 1983. *Introduction to Health Physics,* 2d ed. New York: Pergamon Press. ISBN 0-08-030129-0.

Gollnick, D. A. 1988. *Basic Radiation Protection Technology,* 2d ed. Altadena, Calif.: Pacific Radiation Corp. ISBN 0-916339-03-3.

Shapiro, Jacob. 1981. *Radiation Protection: A Guide for Scientists and Physicians,* 2d ed. Cambridge, Mass.: Harvard University Press. ISBN 0-674-74584-1.

General Reference

Alazraki, Naomi P., and Fred S. Mishkin. 1984. *Fundamentals of Nuclear Medicine.* New York: Society of Nuclear Medicine. ISBN 0-932004-29-6.

Eisenbud, Merril. 1987. *Environmental Radioactivity from Natural, Industrial, and Military Sources,* 3d ed. Orlando, Fla.: Academic Press. ISBN 0-12-235153-3.

L'Annunziata, M. F. 1987. *Radionuclide Tracers: Their Detection and Measurement.* Orlando, Fla.: Academic Press. ISBN 0-124-36252-4.

Lederer, C. Michael, Jack M. Hollander, and Isadore Perlman. 1978. *Table of Isotopes,* 7th ed. New York: Wiley. ISBN 0-47-1041807.

Health Effects from Radiation

Boice, John D., Jr., and Joseph F. Fraumeni, Jr. (eds.). 1984. *Radiation Carcinogenesis: Epidemiology and Biological Significance.* New York: Raven Press. ISBN 0-890049076.

Brill, Bertrand A. (ed.). 1985. *Low-Level Radiation Effects: A Fact Book.*
    New York: Society of Nuclear Medicine. ISBN 0-932004-14-8.
Committee on the Biological Effects of Ionizing Radiations (BEIR). 1980. *The
    Effects on Populations of Exposure to Low Levels of Ionizing Radiation:
    1980.* Washington, D.C.: National Academy Press (BEIR III Report).
    ISBN 0-309-03095-1.
Committee on the Biological Effects of Ionizing Radiations (BEIR). 1988.
    Health Risks of Radon and Other Internally Deposited Alpha-Emitters.
    Washington, D.C.: National Academy Press (BEIR IV Report). ISBN 0-
    309-03789-1.
Committee on the Biological Effects of Ionizing Radiations (BEIR). 1989.
    Health Effects of Exposure to Low Levels of Ionizing Radiation. Wash-
    ington, D.C.: National Academy Press (BEIR V Report). ISBN 0-309-
    03995-9.
National Commission on Radiation Protection and Measurements (NCRP).
    1987. *Recommendations on Limits for Exposure to Ionizing Radiation.*
    NCRP Report No. 91. Bethesda, Md.: NCRP. ISBN 0-913392-89-8.
National Commission on Radiation Protection and Measurements (NCRP).
    1987. *Public Radiation Exposure from Nuclear Power Generation in the
    United States.* NCRP Report No. 92. Bethesda, Md.: NCRP. ISBN 0-
    913392-90-1.
National Commission on Radiation Protection and Measurements (NCRP).
    1987. *Ionizing Radiation Exposure of the Population of the United States.*
    NCRP Report No. 93. Bethesda, Md.: NCRP. ISBN 0-913392-91-X.
National Commission on Radiation Protection and Measurements (NCRP).
    1987. *Exposure of the Population in the United States and Canada from
    Natural Background Radiation.* NCRP Report No. 94. Bethesda, Md.:
    NCRP. ISBN 0-913392-93-6.
National Commission on Radiation Protection and Measurements (NCRP).
    1989. *Comparative Carcinogenicity of Ionizing Radiation and Chemicals.*
    NCRP Report No. 96. Bethesda, Md.: NCRP. ISBN 0-913392-96-0.
National Commission on Radiation Protection and Measurements (NCRP).
    1989. *Exposure of the U.S. Population from Diagnostic Medical Radia-
    tion.* NCRP Report No. 100. Bethesda, Md.: NCRP. ISBN 0-92600-1-0.
National Commission on Radiation Protection and Measurements (NCRP).
    1989. *Exposure of the U.S. Population from Occupational Radiation.*
    NCRP Report No. 101. Bethesda, Md.: NCRP. ISBN 0-929600-05-3.

Waste Disposal

Berlin, R. E., and C. C. Stanton. 1989. *Radioactive Waste Management.*
    New York: Wiley. ISBN 1-85972-0.
Blasewitz, A. G., J. M. Davis, and M. R. Smith (eds.). 1983. *The Treatment
    and Handling of Radioactive Wastes.* Richland, Washington and New
    York: Battelle Press and Springer-Verlag, ISBN 0-935470-14-X.
Burns, Michael E. 1988. *Low-Level Radioactive Waste Regulation: Science,
    Politics and Fear.* Chelsea, Mich.: Lewis Publishers. ISBN 0-87371-
    026-6.

Chapman, Neil A., and Ian G. McKinley. 1987. *The Geological Disposal of Nuclear Waste.* New York: Wiley. ISBN 0-471-91249-2.

EG&G. 1987. *Low-Level Radioactive Waste Management in Medical and Biomedical Research Institutions.* DOE/LLW-13Th. Washington, D.C.: DOE.

Krauskopf, Konrad B. 1988. *Radioactive Waste Disposal and Geology.* New York: Chapman and Hall. ISBN 0-412-28630-0.

League of Women Voters Education Fund. 1982. *A Nuclear Waste Primer.* Washington, D.C.: League of Women Voters. ISBN 0-89959-253-8.

League of Women Voters Education Fund. 1987. *Disposal of Low-Level Radioactive Waste in California.* Pasadena, Calif.: League of Women Voters.

Lipschutz, Ronnie D. 1980. *Radioactive Waste: Politics, Technology, and Risk.* Cambridge, Mass.: Ballinger. ISBN 0-88410-621-7.

Moghissi, A. A., H. W. Godbee, and S. A. Hobart (eds.). 1986. *Radioactive Waste Technology.* Chicago, Ill.: American Nuclear Society.

Murray, Raymond L. 1989. *Understanding Radioactive Waste.* Columbus, Ohio: Battelle Press. ISBN 0-935470-41-7.

National Commission on Radiation Protection and Measurements (NCRP). 1989. *Living Without Landfills.* NCRP Commentary No. 5. Bethesda, Md.: NCRP.

National Low-Level Radioactive Waste Management Program. 1988. *The 1987 State-by-State Assessment of Low-Level Radioactive Wastes Received at Commercial Disposal Sites.* DOE/LLW-69T. Washington, D.C.: DOE.

Resnikoff, Marvin. 1987. *Living Without Landfills.* New York: Radioactive Waste Campaign. ISBN 0-9619078-0-0.

Upton, A. C., T. Kneip, and P. Toniola. 1989. Public Health Aspects of Toxic Chemical Disposal Sites. *Annual Review of Public Health* 10:1–25.

U.S. Congress, Office of Technology Assessment. 1989. *Partnerships Under Pressure: Managing Commercial Low-Level Radioactive Waste.* OTA-0-426. Washington, D.C.: U.S. Government Printing Office.

U.S. Department of Energy. 1988. *Data Base for 1988: Spent Fuel and Radioactive Waste Inventories, Projections, and Characteristics.* DOE/RW-0006, Rev. 4. Washington, D.C.: DOE.

Risk

Covello, V., A., Moghissi, V. Uppuluri, (eds.). 1986. *Uncertainties in Risk Assessment and Management.* New York: Plenum Press.

Hoel, David G., Richard A. Merrill, and Frederica P. Perera (eds.). 1985. *Risk Quantitation and Regulatory Policy.* 19th Banbury Report. New York: Cold Spring Harbor Laboratory. ISBN 0-87969-219-7.

Jasanoff, S. 1986. *Risk Management and Political Culture.* New York: Sage Foundation.

Lowrance, William W. 1976. *Of Acceptable Risk.* Los Altos, Calif.: William Kaufman. ISBN 0-913232-30-0.

Mazur, A. 1987. *The Dynamics of Technological Controversy.* Washington, D.C.: Communications Press.

National Research Council. 1989. *Improving Risk Communication*. Washington, D.C.: National Academy Press. ISBN 0-309-03946-0.

National Safety Council. 1988. *Accident Facts*. Chicago, Ill.: National Safety Council. ISBN 0-87912-139-4.

Nelkin, D. 1987. *Selling Science: How the Press Covers Science and Technology*. New York: Freeman.

Society for Risk Analysis. *Risk Analysis*. Available from Plenum Publishing Corporation, 233 Spring Street, New York, N.Y. 10013.

U.S. Department of Health and Human Services, Centers for Disease Control. *Morbidity and Mortality Weekly Report* Series. Available from the Massachusetts Medical Society, C.S.P.O. Box 9120, Waltham, MA 02254-9120.

Whipple, Chris (ed.). 1987. *De Minimis Risk*. New York: Plenum Press. ISBN 0-306-42530-0.

Federal Laws Pertaining to Hazardous Materials and Radioactive Waste

Clean Air Act (Public Law 91-604)

Clean Water Act (Public Law 92-500)

Comprehensive Environmental Response, Compensation Liability Act (CERCLA) (Public Law 96-510)

Hazardous Materials Transportation Act (Public Law 93-633)

Low-Level Radioactive Waste Policy Act (Public Law 96-573)

Low-Level Radioactive Waste Policy Amendments Act of 1985 (Public Law 99-240)

Marine Protection, Research, and Sanctuary Act of 1972 (Public Law 92-352)

Nuclear Waste Policy Act (Public Law 97-425)

Occupational Safety and Health Act (OSH Act) (Public Law 91-596)

Resource Conservation and Recovery Act (RCRA) (Public Law 94-580)

Safe Drinking Water Act (Public Law 93-523)

Toxic Substances Control Act (TOSCA) (Public Law 94-469)

Ten Frequently Asked Questions about LLRW

1. What is radioactive waste?

Radioactive waste is material resulting from the purification, manufacture, or use of substances containing one or more of certain elements that have unstable atomic structures. Although some of these radioactive elements are man-made, most also occur naturally and have been present for millennia. Their radioactivity, a loss of energy as they are converted to a less energetic, more stable state, is also a function of their radioactive decay to a nonradioactive state. This means that, given time, radioactive waste will become nonradioactive. For some wastes this process will take hours, for most it will take months, and for a small but significant fraction, it will take hundreds or thousands of years.

2. Where is radioactive waste produced?

Low-level radioactive waste results from the purification of radioactive ores, from the manufacture of elements by acceleration of electrons or bombardment by neutrons or protons as caused by fusion or fission, and from the disposal of radioactive pharmaceuticals and other products used for medical diagnostics and treatment or from discarding consumer products such as smoke detectors, gauges, and calibration standards.

3. How is LLRW disposed of?

Dilution and/or decay are the basic approaches for the disposal of LLRW. The present technical solutions are adequate to the task. The decay of the radioactivity is the primary factor to consider when choosing a disposal strategy. Containment need only be long enough to allow the radioactivity to decay, after which the materials will be nonradioactive. The problem inherent in this approach is that a small amount of LLRW may have a very long half-life or decay period, for example, iodine-129 at 17,000,000 years. Since 0.1% of the starting radioactivity will remain after 10 half-lives, significant amounts of this long-lived material must be contained for an essentially indefinite period of time. Dilution with air or water reduces the concentration of the radioactivity but of course not the total amount. If the isotopes occur naturally and the amounts for disposal are an insignificant fraction of the global inventory, radioactive materials can be released at low rates as liquids or gases when the benefits to humankind clearly outweigh any risks. These risks have often been grossly exaggerated in the past without comparison to the commonly accepted risks which we all face daily.

4. Is LLRW dangerous?

Most LLRW does not contain sufficient radioactivity to present a health hazard to those in close proximity to the material. Very few people besides generators and disposal site operators are ever in a position to receive any irradiation from LLRW. The principal potential route of exposure for any member of the public to radioactivity from LLRW would be through ingestion of radioactive materials that have migrated to a water supply. Proper disposal techniques and site monitoring for radioisotope migration should prevent this from happening. No health effects have been observed in populations residing near a radioactive waste disposal facility.

5. Have there been any exposures to the public from radioactivity in LLRW?

None have been measured. However, very low exposures have been estimated at the now closed Maxey Flats disposal facility in Kentucky. The radiation levels detected were far below those permitted for drinking water and no direct evidence has confirmed that the LLRW site was truly the source of this radiation. Subsequent investigations have revealed that the site was not well managed and was the most likely source of contamination and the site was closed. The current LLRW disposal facilities, operating or closed, were constructed before rigorous federal regulations were enacted to improve the shallow land burial of LLRW. In the early years of LLRW disposal, wastes were poorly packaged and unabsorbed liquids were often received. These and other problems have since been addressed by the NRC in 10 CFR Part 61.

6. Why should I be exposed to any radioactivity from LLRW?

Although there are natural sources of radiation to which we are all exposed, a point of view is that all additional exposures should be prevented. In general, the man-made sources of radiation provide only a fraction of the dose received from natural sources. It is noteworthy that medical diagnostics and treatment constitute the single largest category of man-made exposure. The beneficial uses of radioactivity far outweigh the risk they impose. Moreover, despite the political turmoils over the disposal of LLRW, nuclear technologies have been safer than many others heavily used by our society and alternative technologies have not been identified or developed to a practical level and are unlikely to be in the foreseeable future.

7. Is there any level of radioactivity that is safe?

The word "safe" is a relative term. When we say something is safe, we imply that any risks associated with it are very low. It has not been possible to show that health effects are associated with the low levels of radiation found in LLRW. This does not mean that such effects do not take place, only that they occur so infrequently that they cannot be distinguished from health effects from many other causes. It is also possible that below a certain level of radioactivity, no health effects will be observed, either because they never take place or more likely because mechanisms at the molecular level of our genes operate continuously to repair damage from the many naturally oc-curring mutagens (and carcinogens) to which we are constantly exposed. Indeed, many individuals have been either acutely or chronically exposed to levels several times higher than the average annual levels of radioactivity absorbed without any adverse health effects observed. The overwhelming majority of scientific opinion considers that any potential exposure to the radioactivity from LLRW is safe and that regulations to minimize even that exposure are conservative and effective.

8. What assurance do we have that LLRW will be managed properly?

From review of the current legislation and regulations affecting LLRW, it appears that any of the disposal strategies under consideration will be adequate to protect people from exposure to LLRW. However, continued vigilance will be required to assure that these regulations are followed in the years to come. The present public concern over waste disposal also reflects the concern of those generating LLRW. It is likely that adequate inspection and environmental monitoring will be performed by all parties concerned as long as these issues remain a priority. From a long-term perspective, no assurances can really be given for the small amount of LLRW whose activity is truly long-lived. From all information available, it appears that these materials will be sufficiently contained to permit their long-term decay. By segregating these materials from those with shorter half-lived nuclides, it will be easier for future generations to continue to evaluate the adequacy of today's technology. In this regard all of the by-products of our technological society will have to be studied by future generations, but we must act as responsibly as our knowledge permits, and we must act now.

9. Why must LLRW be buried in anyone's backyard, least of all mine?

Regardless of what types of volume reduction methods or treatments we subject LLRW to, some amount will have to be buried, whether above or below ground or in simple or sophisticated structures. Given that migration through water has been the major route for isotopic releases affecting people in the surrounding area, the location of the burial site must be dictated by the geophysical properties of the site. Strong emphasis must be placed on volume reduction and pretreatment. The fewer the number of facilities needed, the better, as these facilities will carry with them both financial and environmental liabilities.

10. Who is paying for LLRW disposal?

The direct answer is that the generator is paying for disposal. The generator is responsible for any impacts from this material "from cradle to grave," and therefore has a vested interest in knowing that the LLRW has been properly disposed of. So far the generators who have borne most of the costs associated with the development of plans and sites for LLRW disposal have been utilities with nuclear power reactors. Since the benefits are widely distributed, the costs for LLRW disposal are passed on to all of us, the consumers of the many products and services utilizing radioactive materials.

Key Regulatory Agencies, Advisory Groups, and Compacts

American Conference of Governmental Industrial Hygienists
6500 Glenway Avenue
Building D-7
Cincinnati, OH 45211-4438
(513) 661-7881

American Nuclear Society
555 North Kensington Avenue
La Grange Park, IL 60525
(312) 352-6611

ANDRA (Agence Nationale pour la Gestion des Dechets Radioactifs)
31-33, rue de la Federation
75015 Paris, France
40-56-19-00

Atomic Energy Control Board
Waste Management Division
270 Albert Street
P.O. Box 1046
Ottawa, Ontario
K1P 5S9 Canada
(613) 995-4055

Atomic Industrial Forum, Inc.
7101 Wisconsin Avenue
Bethesda, MD 20814-4891
(301) 654-9260

California Department of Health Services
Radiological Health Division
1449 West Temple Street
Los Angeles, CA 90026
(213) 620-2860

Center for Devices & Radiological Health
United States Food and Drug Administration
5600 Fishers Lane
Rockville, MD 20857
(301) 443-1544

Central Midwest Compact Commission
Director
Office of Energy Research
901 S. Matthews Avenue
Urbana, IL 61801
(217) 333-7734

Central States Compact Commission
Golds Galleria
1033 O Street, Suite 625
Lincoln, NE 68508
(402) 476-8247

Colorado Department of Health
Radiation Control Division
4210 E. 11th Avenue
Denver, CO 80220
(303) 320-8333

Connecticut Hazardous Waste Management Service
900 Asylum Avenue
Suite 360
Hartford, CT 06105-1904
(203) 244-2007

Electric Power Research Institute (EPRI)
3412 Hillview Avenue
Palo Alto, CA 94005
(415) 855-2089

Health Physics Society
8000 Westpark Drive, Suite 400
McLean, VA 22102
(804) 790-1745

Illinois Department of Nuclear Safety
1035 Outer Park Drive
Springfield, IL 62704
(217) 785-9868

International Atomic Energy Agency
P.O. Box 100
A-1400 Vienna, Austria
011-43-1-23600

International Commission on Radiation Units and Measurements
7910 Woodmont Avenue
Bethesda, MD 20814
(301) 657-2652

International Commission on Radiological Protection
Clifton Avenue
Sutton, Surrey SM2 5PU
United Kingdom

Kentucky Department of Environmental Protection
18 Reilly Road
Frankfurt, KY 40601
(502) 564-3035

League of Women Voters Education Fund
Energy Department
1730 M Street, N.W.
Washington, DC 20036
(202) 296-1770

Maine Low-Level Radioactive Waste Authority
99 Western Avenue
Augusta, ME 04330
(207) 626-3249

Massachusetts Low-Level Radioactive Waste Management Board
100 Cambridge Street
Boston, MA 02202
(617) 292-5589

Michigan Low-Level Radioactive Waste Authority
P.O. Box 30026
Lansing, MI 48909
(517) 335-0437

Midwest Compact Commission
350 N. Robert Street
St. Paul, MN 55101
(612) 293-0126

National Academy of Science
Radioactive Waste Management Board
2101 Constitution Avenue NW
Washington, DC 20148
(202) 334-3068

National Council on Radiation Protection and Measurements
7910 Woodmont Avenue
Bethesda, MD 20814
(301) 657-2652

National Technical Information Service
U.S. Department of Commerce
5285 Port Royal Road
Springfield, VA 22161
(804) 487-4660

Nebraska Environmental Control
301 Centennial Mall, South
P.O. Box 98922, Statehouse Station
Lincoln, NE 68509-8922
(402) 471-2244

New Jersey Low-Level Radioactive Waste Facility Siting Board
Building CN415
Trenton, NJ 08625-0415

New York State Department of Environmental Conservation
50 Wolf Road
Albany, NY 12233
(518) 457-5915

New York State Energy Research and Development Authority
Empire State Plaza, Albany Building
Albany, NY 12223
(518) 432-1400

New York State Low-Level Radioactive Waste Siting Commission
2 Third Street, Fourth Floor
Albany, NY 12180
(518) 271-1585

North Carolina Low-Level Radioactive Waste Management Authority
116 West Jones Street, Suite 2109E
Raleigh, NC 27603
(919) 733-0682

Northeast Compact Commission
195 Nassau Street
Princeton, NJ 08542
(609) 497-1447

Northwest Compact Commission
Washington State Department of Ecology
Low-Level Radioactive Waste Program
Mailstop PV-11
Olympia, WA 98504-8711
(206) 459-6244

Pennsylvania Department of Environmental Resources
Bureau of Radiation Protection
16th Floor, Fulton Building
P.O. Box 2063
Harrisburg, PA 17120
(717) 787-2163

Rocky Mountain Compact Board
1675 Broadway, Suite 1400
Denver, CO 80202
(303) 825-1912

Society of Nuclear Medicine
136 Madison Avenue
New York, NY 10016
(212) 889-0717

Southeast Compact Commission
Suite 100B
3901 Barrett Drive
Raleigh, NC 27609
(919) 781-7152

Texas Low-Level Radioactive Waste Disposal Authority
Suite 300
7703 North Lamar
Austin, TX 78752
(512) 451-5292

Union of Concerned Scientists
26 Church Street
Cambridge, MA 02238
(617) 547-5552

United Kingdom Atomic Energy Authority
AEE Winfrith
Dorchester, Dorset, England

United Nations Scientific Committee on the Effects of Atomic Radiation
Vienna International Centre
P.O. Box 500
A-1400 Vienna, Austria

United States Department of Energy
Office of Nuclear Waste Management
Mailstop B107, Germantown Building
Germantown, MD 20545
(301) 353-5645

United States Department of Health and Human Services
National Institute for Occupational Safety and Health
4676 Columbia Parkway
Cincinnati, OH 45226
(513) 533-8236

United States Department of Labor
Occupational Safety and Health Administration
Washington, DC 20210
(202) 523-8148

United States Department of Transportation
Materials Transportation Bureau—Office of Hazardous Materials Operations
400 7th Street, S.W.
Washington, DC 20590
(202) 426-4000

United States Environmental Protection Agency/Office of Radiation Programs
401 M Street S.W.
Washington, DC 20460
(202) 557-9380

United States Food and Drug Administration
5600 Fishers Lane
Rockville, MD 20857
(301) 443-1544

United States Nuclear Regulatory Commission
1717 H Street N.W.
Washington, DC 20555
(202) 634-7511

Vermont Advisory Commission on Low-Level Radioactive Waste
Center Building
103 S. Main Street
Waterbury, VT 05676
(802) 244-5164

Washington Department of Ecology/Low-Level Radioactive Waste Program
Mailstop PV-11
Olympia, WA 98504-8711
(206) 459-6863

Wisconsin Radioactive Waste Review Board
3 South Pinckney Street
Madison, WI 53704
(608) 266-9810

Acronyms

AEC	Atomic Energy Commission
ALARA	As low as reasonably achievable
BEIR	Biological effects of ionizing radiation
BNL	Brookhaven National Laboratory, Brookhaven, N.Y.
BRC	Below regulatory concern
CERCLA	Comprehensive Environmental Response, Compensation and Liability Act (Superfund)
DOE	Department of Energy
DOL	Department of Labor

DOT	Department of Transportation
EIS	Environmental Impact Statement
EPA	Enviromental Protection Agency
EPRI	Electric Power Research Institute
FDA	Food and Drug Administration
GTCC	Greater than Class C radioactive waste
HHS	Department of Health and Human Services
HIC	High integrity container
HLW	High-level (radioactive) waste
IAEA	International Atomic Energy Agency
IAEC	International Atomic Energy Commission
ICRP	International Commission of Radiological Protection
ICRU	International Commission on Radiation Units
LLRW	Low-level radioactive waste
LLRWPA	Low-Level Radioactive Waste Policy Act of 1980
LLRWPAA	Low-Level Radioactive Waste Policy Amendments Act of 1985
MESA	Mining Enforcement and Safety Administration
MPC	Maximum permissible concentration
MRS	Monitored retrievable storage
NARM	Naturally occurring and accelerator produced radioactive materials
NCRP	National Council on Radiation Protection and Measurements
NEPA	National Environmental Policy Act
NIMBY	"Not in my backyard"
NRC	Nuclear Regulatory Commission
NTIS	National Technical Information Service, Springfield, Va.
NYSDEC	New York State Department of Environmental Conservation
NYSERDA	New York State Energy Research and Development Authority
OSHA	Occupation Safety and Health Administration
RCRA	Resource Conservation and Recovery Act
RO	Reverse osmosis
SI	International System of Units *(French: System Internationale)*
SLB	Shallow land burial
SNL	Sandia National Laboratory, Albuquerque, N.M.
TRU	Transuranic
TSCA	Toxic Substance Control Act
UMTRAP	Uranium Mill Tailings Remedial Action Program
UNSCEAR	United Nations Scientific Committee on the Effects of Atomic Radiation
WIPP	Waste Isolation Pilot Plant, Carlsbad, N.M.
WNYNSC	Western New York Nuclear Services Center
WVDP	West Valley Demonstration Project

GLOSSARY

Absorbed dose Amount of energy imparted to a mass. Traditional unit is the *rad*, roughly equivalent to the amount of energy deposited in tissue by one *roentgen* of radiation exposure. SI unit is the gray (Gy); 1 Gy = 100 rads.

Accelerator Device used to increase the velocity and energy of charged particles.

Actinides Elements of the seventh series in the Periodic Table. All members of this series have an atomic mass greater than actinium and are radioactive. Examples include thorium, uranium, plutonium, and americium.

Activation products Elements that become radioactive during the course of bombardment by neutrons or protons.

Agreement state Any state that has entered into an agreement with the Nuclear Regulatory Commission, as specified in the Atomic Energy Act of 1954, so the state is granted the authority to regulate radioactive materials.

Alpha-emitter An element that in undergoing radioactive decay releases alpha particles (helium nuclei).

Alpha particle A form of radiation consisting of a helium nucleus. Helium contains two protons and two neutrons but no encircling electrons. Alpha particles are the largest of the radioactive particles and are easily stopped. Taken internally, however, they can do great damage because of their large size.

Ampoule A sealed glass or plastic vial used for storage of small amounts of liquid, solid, or gas.

Aquifer Permeable layers of subsurface gravel or sand that contain groundwater. The strata overlying an aquifer are of great importance as they will determine the potential for contaminants released on the surface to migrate into the water table. Confined aquifers are bounded by impermeable strata. Although aquifers generally receive discharge from surface waters, the rate of water withdrawal for human uses (agricultural, industrial, and domestic consumption) often exceeds the natural rate of recharge. Such a condition literally "mines" this important water resource. Siting regulations for the

191

disposal of LLRW strictly prohibit the development of a site over a primary drinking water aquifer.

As low as reasonably achievable (ALARA)   Basic principle of radiation protection, stating that doses from radioactive materials should be reduced to the lowest possible levels, provided that economic and social benefits exceed any risks.

Atom   The smallest particle of an element that retains all the chemical and physical characteristics of that element.

Backfill   The placement of soil or other material in, around, or over a structure.

Background radiation   Radiation that arises from constant natural sources and accepted man-made sources, such as dental and medical x-rays. Radiation from cosmic sources and natural radiation are always present. Approximately 55% of background radiation is due to radon emitted from rock and building materials.

Becquerel (Bq)   The international unit for radioactivity representing one disintegration per second. Named in honor of Henri Becquerel, who discovered the property of *radioactivity* in 1896. By international agreement, the becquerel has recently replaced the *curie* as the preferred unit of measure. One Bq = 2.7×10^{-11} Ci.

Below Regulatory Concern (BRC)   Regulatory concept for low levels of radioactivity, dependent upon the waste stream, disposal strategy, and the risk-societal benefit.

Beta-emitter   An element that releases beta particles (negative electrons or positrons) during radioactive decay.

Beta particle   An electron (or positron) emitted by a radionuclide. Electrons have low mass, a net negative charge, and travel with great speed, making them much more penetrating than alpha particles.

Biogeochemical cycles   Processes through which chemicals such as carbon, oxygen, phosphorus, nitrogen, sulfur, and water are cycled within the biosphere. These systems are complex and require the interplay between physical and biological factors.

Biological half-life   Time required for one-half of the amount of a chemical or radionuclide to be eliminated from a living organism, dependent upon the body's affinity for the chemical, its solubility, retention, metabolism, and rate of excretion.

Biosphere   Those parts of the earth and its atmosphere that support life. It represents only about 11 km of the atmosphere, the contents of the earth's surface, and a few kilometers into the earth's interior.

Breeder reactor   A nuclear reactor design that produces more nuclear fuel than it consumes. The process converts nonfissile ^{238}U into fissionable ^{239}Pu by neutron bombardment. It is a useful method for extending the supply of nuclear fuel since ^{238}U is much more abundant than the ^{235}U typically used in a fission reactor. However, great increases in the world's inventory of ^{239}Pu are not attractive, as it is highly carcinogenic and could be used to fashion crude nuclear weapons if stolen by terrorists. No commercial breeder reactors operate in the United States today.

By-product material Radioactive material produced in reactors by irradiation or through the extraction of uranium from ore.

Carcinogen Any chemical or physical agent (e.g., radiation) that causes cancer.

Cladding Corrosion-resistant tube, usually made of zirconium alloy or stainless steel, which surrounds the fuel pellets used in the core of a nuclear reactor. The cladding provides protection from a chemically reactive environment and contains the fission products.

Code of Federal Regulations Documentation of the general rules by the executive departments of the federal government. The code is divided into 50 titles that represent broad areas subject to federal regulation. Each title is divided into chapters that usually bear the name of the issuing agency. Each chapter is further divided into parts covering specific regulatory areas.

Collective dose See *person-rem.*

Compact An association of states, ratified by the U.S. Congress, created in response to the mandates of the 1980 Low-Level Radioactive Waste Policy Act and the 1985 Low-Level Radioactive Waste Policy Amendments Act in order for the involved states to share the responsibilities of establishing an LLRW disposal site.

Compaction Physical compression of waste to reduce its volume, usually accomplished with a hydraulic press. The force of compaction can vary from a few hundred pounds per square inch to upwards of 500,000 psi.

Concentration The amount of radioactivity in a given mass of material, for example, $\mu Ci/g$ or Bq/g.

Cosmic rays Radiations originating beyond the earth's atmosphere. The primary cosmic ray particles are mainly protons, which interact with atoms in the atmosphere to produce a wide variety of secondary particles. The dose from cosmic rays is produced by these secondary particles.

Critical mass The amount of fissile material (e.g., ^{235}U) required to initiate a highly energetic chain reaction via neutron bombardment.

Curie (Ci)   The traditional unit for radioactivity named after Mme. Curie for her pioneering work with radium a century ago. One curie is defined as the amount of radioactive material which produces 3.7×10^{10} nuclear disintegrations per second, approximately the activity contained in one gram of radium. Worldwide scientific adoption of a standard system of units now uses the unit *becquerel (Bq)*; 1 Ci = 3.7×10^{10} Bq.

Cyclotron   A type of accelerator that propels charged particles, such as protons or ions, using an alternating electric field and a constant magnetic field.

Daughter nuclides   The intermediate isotopes produced during a decay chain.

Decay (radioactive)   The transition of a nucleus from one energy level to another, usually accompanied by the emission of a photon, electron, or neutron.

Decay chain   A sequential series of radioactive emissions that ultimately transforms a radionuclide into a stable, nonradioactive element.

Decommissioning   The process of removing a former radioactive facility or area from operation and decontaminating and/or disposing of it or placing it in a condition of standby with appropriate controls and safeguards.

Decontamination   Removal of radioactive material from the surface or interior of another material.

de minimis   Abbreviation for *de minimis non curat lex,* "the law does not pay attention to the trivial." For radiation protection purposes, this term defines an amount of radioactivity considered to be trivial and that should not be regulated regardless of the number of people exposed or the duration of exposure.

DNA   Deoxyribonucleic acid, the fundamental chemical that determines the genetic code. Short sequences of DNA comprise genes, the basic units of inheritance. Our chromosomes are composed of hundreds of thousands of genes.

DOE/defense waste   Radioactive waste produced from activities supported by the Department of Energy and/or defense programs of the U.S. government.

Dose   A measurement of the quantity of radiation or energy absorbed per unit mass (m*rem*/year or m*sievert*/year).

Dose-response   The relationship between the dose of a chemical or physical agent and its corresponding effect.

Dosimeter   A device that measures doses of radiation.

Drum   An empty cylinder used for packing, storing, and disposing of radioactive waste. Typically, drums are made of carbon steel and have a capacity of 55 gal.

Effective annual dose equivalent   The amount of absorbed dose received by a specific tissue or organ in a year. The total effective annual dose equivalent is the product of the total absorbed dose and tissue- or organ-specific weighting factors defined by the ICRP. This dose is measured in **sieverts.**

Effective half-life   The time required for one-half of the amount of radionuclide to be eliminated from a living organism, dependent upon both the physical and biological half-life of the nuclide.

Effluent   Liquid or gaseous waste materials released to the biosphere from a distinct point source.

Electromagnetic radiation   A broad spectrum of radiation consisting of electric and magnetic waves that travel at the speed of light, including radio waves, gamma waves, and x-rays.

Electron   Fundamental physical particle that possesses one unit of negative charge but which has little resting mass. In the Bohr model of the atom, electrons circle a positively charged nucleus at great speed. In stable, nonradioactive atoms, the net negative charge of the electrons is balanced by the sum of the positive charges of the protons in the nucleus. In most radionuclides, neutrons outnumber protons.

Electron capture (EC)   A mode of radioactive decay that results in an emission of photons similar to gamma radiation.

Element   One of the 106 known types of substances that comprise, at or above the atomic level, all matter. The number of protons in an atom of an element is equal to its atomic number and determines its position within the Periodic Table. All elements with atomic numbers higher than lead are radioactive. Elements with atomic numbers higher than uranium (the transuranics) are created by bombardment of other elements with neutrons or other heavier particles.

Emergency core-cooling system   A system of last resort incorporated in most nuclear reactors. It is designed to flood the reactor core with large volumes of water to cool the fuel rod assemblies in the event that an uncontrollable critical mass develops and to prevent a reactor core "meltdown."

Enrichment   The process of artificially increasing the abundance of naturally occurring radioactive material. Usually mentioned in connection with the nuclear fuel ^{235}U, which must be enriched relative to the more prevalent, but nonfissionable, uranium isotope ^{238}U.

Environmental impact statement (EIS)   A report that documents the information required to evaluate the environmental impact of a project. Such a report informs decision makers and the public of the reasonable alternatives that would avoid or minimize adverse impacts or enhance the quality of the environment. The National Environmental Policy Act requires that all federally funded projects that might have deleterious effects on the environment be subjected to review through an EIS.

Epidemiology   The branch of medical science that studies the incidence, distribution, and control of diseases.

Exempt quantity   An amount of radioactive material that may be used without regulation.

Exposure   A measure of ionization produced in air by x- or gamma-rays. Unlike dose, exposure refers to a potential for receiving radiation. Acute exposure generally means a high level of exposure over a short time period, whereas chronic exposure refers to low levels of exposure over a long period of time. Traditional unit is the *roentgen*; SI unit is coulombs per kilogram.

Fallout   The descent to the earth's surface of particles contaminated with radioactive material primarily from the testing of atomic weapons, but also from radioactive materials released from nuclear reactors, processing plants, and some natural sources such as volcanoes and large fires. This term also applies to the contaminated particles themselves. Particles that return to the earth's surface within 24 hours are called local or early fallout. Worldwide or delayed fallout is brought down from the upper troposphere and stratosphere by rain and snow over extended time periods.

Fissile material   Radioactive material capable of achieving and maintaining fission; for example, ^{235}U, ^{239}Pu.

Fission   The splitting of a heavy nucleus into two approximately equal nuclei of lighter elements. This process is accompanied by the release of energy and generally one or more neutrons. Although, at least in theory, fission could occur naturally, it is usually initiated by the purposeful bombardment of nuclear fuel with gamma rays, neutrons, or other particles.

Fuel cycle   The series of steps involved in supplying fuel for nuclear power reactors. It includes the mining and refining of fuel elements, their use in a reactor, chemical processing to recover fissionable material remaining in the spent fuel, reenrichment and refabrication, transportation, and management of the resulting radioactive waste.

Fuel reprocessing   Recovering uranium and plutonium for reuse from irradiated (spent) nuclear reactor fuel.

Fusion   The union of atomic nuclei to form heavier nuclei, accompanied by the release of enormous amounts of energy.

Gamma-emitter   A radioisotope that produces energy in the form of gamma rays as it decays. Also included in this category are *x-rays* and *electron capture*.

Gamma rays   Very short wavelength electromagnetic radiation. Gamma rays have no mass and travel with high frequency, making them highly penetrating.

Genetic effects   Radiation-induced changes in the genetic material of an organism (i.e., germ-line cells of the ova and spermatozoa).

Gray (Gy)   The international unit for absorbed dose; 1 Gy = 100 rad.

Greater than Class C (GTCC)   Waste from commercial sources containing radionuclide concentrations that exceed the NRC's limits for Class C LLRW, as defined in 10 CFR Part 61.55.

Half-life   The time required for one-half of the atoms of a particular radionuclide to decay. After a period of time equal to 10 half-lives, the radioactivity remaining is only 0.1% of the original amount present.

Hazard   A potential source of danger or harm. Contrast with *risk*.

Heavy metals   Metallic elements with relatively high atomic mass such as cadmium, lead, mercury, and nickel. All are toxic and several (e.g., cadmium, nickel) are known carcinogens.

HEPA filter   High-efficiency particulate aerosol filter. A filter capable of removing more than 99.97% of airborne particles that exceed 0.3 μm diameter.

Hereditary effects   *Stochastic* effects in the progeny of individuals who were exposed to radiation.

High-level waste (HLW)   As defined by the Nuclear Waste Policy Act, high-level waste is (1) the highly radioactive material resulting from the reprocessing of spent nuclear fuel, including the liquid waste produced directly in reprocessing and any solid material derived from such liquid waste that contains fission products in sufficient concentrations; and (2) other highly radioactive material that the

NRC, consistent with existing law, determines by rule to require permanent isolation.

Hydrocarbons   Organic molecules composed of hydrogen and carbon.

Inadvertent intrusion   Violation of a closed waste facility due to ignorance of its location and/or contents. Intrusion may result, for example, from digging for natural resources or agricultural development in an area formerly used for waste disposal.

Incineration   A controlled combustion by which wastes are burned and converted into CO_2 and H_2O. Incomplete combustion, as can occur at low temperatures or short residence time, will prevent the full destruction of the waste and may release various intermediate compounds as well as some of the original material. Incineration of hazardous waste generally occurs at high temperatures. The use of cyclones, baghouses, and scrubbers greatly improves the cleanliness of the effluent.

Ion   An atom or molecule that has gained or lost electrons, yielding a net electrical charge (+ or −) on the unit. The process of gaining or losing electrons is called ionization.

Ionizing radiation   Any radiation that produces ions by displacing electrons from atoms or molecules.

Isotope   One of two or more atoms with the same atomic number but different atomic weights. Isotopes of a given element contain the same number of protons but different numbers of neutrons.

Landfill   A land site for waste disposal that lacks any provisions for protecting ground and surface waters from pollution. Contrast with *sanitary landfill* and *secured landfill*.

Latency   The time lag between exposure to a potentially harmful agent such as radiation and the appearance of health effects.

Leachate   The solution resulting from the extraction of a soluble component of a solid by a solvent, usually water, percolating through the solid.

Lead pig   A container made of lead expressly for shielding radiation.

Leukemia   Cancer of the white blood cells (leukocytes) and their precursor cells. For epidemiological purposes, it is a useful "indicator" of past exposure to certain carcinogens since leukemia has a relatively short (less than 10 years) latency period.

Long-lived radioisotope   A radioisotope with a relatively long half-life. With respect to LLRW, radioisotopes with half-lives longer than 5 years are considered to be long-lived.

Low-level radioactive waste (LLRW)   Radioactive waste not classified as high-level waste, transuranic waste, spent nuclear fuel, or by-product material such as uranium or thorium tailings and wastes.

Manifest   A legal document that describes the contents of a shipment of hazardous or radioactive waste. It originates with the generator and is completed at the disposal site, permitting the tracking of the waste at all points.

Maximum permissible concentration (MPC)   The amount of radioactive material in air, water, or food that might be expected to result in a *maximum permissible dose* to persons continuously exposed and internalizing them at a standard rate of intake, and from which no appreciable health damage should occur.

Maximum permissible dose (MPD)   The dose of ionizing radiation below which there is no reasonable expectation of risk to human health and that, at the same time, is below the lowest level at which a definite risk is known to exist.

Mined cavities   See *repository, geologic.*

Mixed waste   Waste contaminated with both radioactivity and hazardous chemicals that is regulated by both the NRC and EPA or their local counterparts.

Monitored retrievable storage (MRS)   Concept of a temporary storage facility for high-level radioactive wastes and spent fuel. Congress has linked construction of MRS to licensing of the Yucca Mountain spent-fuel and HLW repository.

Mutation   A change in the genetic material (**DNA**). Mutations arise spontaneously and act as a driving force behind evolution. The rate of mutation may be increased by exposure to certain chemical and physical agents.

NARM   Naturally occurring or accelerator-produced radioactive material.

Neutron   A fundamental nuclear particle that carries no electrical charge. Neutrons are much heavier than electrons. Their high speed and unpredictable travel course give neutrons a greater damage potential than gamma rays or beta particles.

NIMBY   Not in my backyard. A colloquial phrase used to characterize the tendency of people to oppose the siting of a new facility, for example, a prison, an incinerator, or a low-level radioactive waste disposal facility, near their residences for fear of health effects and/or loss of property value.

Nuclear reactor   A device in which a controlled nuclear chain reaction is maintained, either for the purpose of experimentation, production of weapons-grade fissionable material, or the generation of electricity.

Nuclide A specific atom characterized by its atomic mass, which indicates the number of protons and neutrons in its nucleus.

Occupational exposure Exposure received during the course of employment.

Percolation The downward movement of liquids through a permeable medium.

Person-rem (person-sievert) Epidemiological concept for quantitating cumulative and collective doses of radiation. It is the sum of the radiation dose received by the entire population of a given area. All conditions equal, the larger the population defined, the larger the collective dose.

pH A numeric scale that indicates the hydrogen ion concentration of a substance. The scale is logarithmic and reads from 0 (acid) to 14 (basic). The pH of a liquid is an important property as it will influence leaching, corrosion, and chemical reactions with the surroundings.

Proton A fundamental nuclear particle that is positively charged and has a mass about equal to a neutron. Together with neutrons, these particles comprise an atomic nucleus. Protons have penetration abilities similar to electrons.

Rad Radiation absorbed dose. Unit of absorbed dose of radiation representing the absorption of 100 ergs of ionizing radiation per gram of absorbing material; 1 rad = 0.01 Gy.

Radiation Energy in the form of waves or particles.

Radioactive contamination Amounts of radioactive material remaining on the surface or in another material following contact with a radioactive substance.

Radioactivity The spontaneous emission of radiation from unstable atomic nuclei. Emissions take the form of *alpha particles, beta particles, gamma/x-rays, electron capture,* or *neutrons* or a mixture of these. The amount of radioactivity present was traditionally measured in *curies,* but by international convention the unit *becquerel* is now used.

Radiochemical Any chemical containing radioactive atoms.

Radionuclide A radioactive isotope, also called radioisotope.

Radon A naturally occurring radioactive gas that is a decay product of radium-226. Radon is released from the decay of radium present in soil, rock, and some building materials. Radon-222 is the most common and longest lived (half-life = 3.82 days) of the radon isotopes, all of which are alpha-emitters.

Reactor, boiling water (BWR)   A light-water reactor in which water, used as both coolant and moderator, is boiled in the core. The resulting steam is used directly to drive a turbine generator.

Reactor, heavy-water   A nuclear reactor that uses "heavy" (D_2O) water as the primary coolant and moderator, with slightly enriched uranium as fuel.

Reactor, light-water   A nuclear reactor that uses "light" (regular, H_2O) water as the primary coolant and moderator, with slightly enriched uranium as the fuel. The two types of commercial light-water reactors are boiling-water and pressurized-water.

Reactor, pressurized-water (PWR)   A light-water reactor in which heat is transferred from the core to a heat exchanger via water kept under high pressure, so that high temperature can be maintained in the primary system without boiling the water. Steam is generated in a secondary circuit and used to drive a turbine.

Reactor, research   A reactor whose nuclear radiations are used primarily as a tool for basic or applied research. Typically, it has a thermal power of 10 MWt or less and may include facilities for testing reactor materials.

Relative biological effectiveness   Given the same absorbed dose, the different ionizing radiations produce biological effects that consistently differ in their severity. The absorbed dose from different radiations can be compared on the same scale by applying modifying factors called quality factors to account for their differential effects in tissue.

rem   Roentgen equivalent man. Traditional unit for dose equivalent. Equal to the radiation absorbed dose times a quality factor based on the *relative biological effectiveness* of the radiation. One rem equals approximately one rad for gamma and x-rays and most beta particles, 0.5 to 0.1 rad for neutrons, and approximately .05 rad for alpha particles. 1 rem = 0.01 Sv.

Repository, geological   A facility with an excavated subsurface system used for the permanent disposal of spent fuel and high-level radioactive waste.

Risk   Hazard with a quantifiable probability of occurrence.

Roentgen (r)   Traditional unit for exposure to gamma or x-ray radiation. One roentgen produces an absorbed dose in tissue of approximately one rad.

Sanitary landfill   A waste disposal facility employing trench or pit burial. Most municipal waste is disposed of in sanitary landfills.

Scintillation fluids Liquid organic compounds used to enhance the detection and quantification of weak-energy beta particles.

Sealed source A self-contained, encapsulated source of radioactivity.

Secured landfill A waste disposal method similar to a sanitary landfill but which has been sited, designed, and operated with greater stringency. Secured landfills should be located in areas with soils of low permeability and lined with materials designed to trap and collect any leachate. Such a facility must also have monitoring systems in place to detect the migration or release of hazardous substances. Secured landfills are usually reserved for hazardous chemical disposal, although some of their features will be incorporated in the new LLRW facilities due to come on-line in the 1990s. A properly designed shallow land burial facility is an example of a secured landfill used for the disposal of low-level radioactive waste.

Shallow land burial The disposal of wastes within 30 meters of the earth's surface, covered with soil and other overburden materials.

Short-lived radioisotopes Radioisotopes with relatively short half-lives. With regard to LLRW, radioisotopes with half-lives less than 5 years are considered to be short-lived.

Sievert (Sv) The international unit for dose equivalent. One sievert equals an absorbed dose of one joule per kilogram; 1 Sv = 100 rem.

Somatic effects Changes in the somatic (nongermline) cells of an organism. Compare with *hereditary* effects.

Source material Uranium, thorium, or ore containing 0.05% of either element, which is used as fissionable material.

Special nuclear material Radioactive material that is enriched in uranium-235 or man-made fissile elements, such as plutonium, used in both nuclear power reactors and weapons.

Specific activity The amount of radioactivity per amount of material, usually measured in curies per mmol or becquerels per mmol.

Spent fuel Nuclear fuel that has been permanently discharged from a reactor after it has been irradiated. Typically, spent fuel is measured in terms of either the number of fuel assemblies discharged or the mass of the discharged fuel.

Stochastic effects The dose-dependent probability that an event will occur, independent of its severity.

Subsidence The collapse or downward sinking of the soil surface or, on a larger scale, the earth's crust.

Superfund The Comprehensive Environmental Response, Compensation and Liability Act of 1980 (CERCLA). This act provides federal monies for the remediation of toxic waste sites.

Synergistic   The effect of two or more factors in combination is greater than the simple sum of each individual factor, such as the action of asbestos and smoking in producing lung cancer.

Tailings   Residual slurry of sand and clay particles that results from the mining and milling of uranium ore.

Teratogen   A substance that causes malformation of the fetus if a female is exposed to it during pregnancy.

Threshold effect   The concept that no response will occur unless an exposure exceeds some critical concentration, the threshold value. Radiation is generally regulated as if no threshold exists. Proponents of the threshold effect argue that the body has defense mechanisms capable of detoxifying many chemicals and of repairing DNA damage such as that which occurs from low levels of radiation.

Transuranic (TRU) waste   Radioactive waste that contains more than 100 nCi/g (3700 Bq) of alpha-emitting isotopes with atomic numbers greater than 92.

Toxic   Poisonous.

Vitrification   Encapsulation in glass.

Volume reduction   Treatment of existing waste materials in order to decrease the amount of waste that must be shipped to a waste disposal facility.

Waste minimization   Elimination or modification of practices that result in waste in order to reduce the total amount of waste produced.

Waste stream   A specific type of waste, characterized by its physical and chemical properties as well as the process(es) that generates it. A single generator may produce several waste streams, for example, a solid waste stream of used protective equipment and glassware and a liquid waste stream of contaminated cooling water.

X-rays   Penetrating electromagnetic radiation having a wavelength much shorter than that of visible light.

INDEX

Waste generators x, 19, 36, 38–40, 86,
133, 147, 169, 170
non-utilities (or non-fuel-cycle) 57,
67, 68
governmental x, 19, 25–27, 38–40
industrial x, 19, 22–24, 38–40, 62,
76, 91, 169, 170, 171
institutional x, 19, 22, 24–25, 31,
32, 37–40, 76, 91, 169, 170,
171
utilities (or nuclear power) x, 19,
20–22, 32, 37–40, 61, 67, 68,
76, 84, 87, 169, 170, 171
Water treatment 85, 113, 114
Weapons testing xii, 147
Western New York Nuclear Services
Center (WNYNSC) 51

West Germany. (*See* Federal Republic
of Germany)
Westinghouse 24, 50, 62, 133
West Valley, New York. (*See* Disposal
sites)
West Valley Demonstration Project
(WVDP). (*See* Disposal sites:
West Valley)
West Valley Nuclear Services Com-
pany 53

X-rays xiii, 148, 153, 203

Yucca Mountain 7, 111